On the cover: (front) *Mer-veille* (Museum of European and Mediterranean Civilizations) by Yann Kersalé and architect Rudy Ricciotti in Marseille (see p. 24), photo © 2015 Lisa Ricciotti - R. Ricciotti et R. Carta architectes - light installation Yann Kersalé-SNAIK/MuCEM; (back) *1.26* by Janet Echelman (see p. 146).

p.1 *Il Giardino Verticale* by Richi Ferrero in the courtyard of Turin's Palazzo Valperga Galleani (details p.264).

p.4 The world's largest Chinese lantern sculpture, by CL3 and LEDARTIST, at a festival in Hong Kong (details p.264).

pp.6–7 *Pixel Cloud* light projection and sound show, by UNSTABLE, at Reykjavik's 2013 'Winter Lights' festival (details p.264).

Superlux: Smart Light Art, Design and Architecture for Cities © 2015 Davina Jackson

Texts © 2015 their respective authors

Designed by Deuce Design

First published in 2015 in hardcover in the United States of America by Thames & Hudson Inc., 500 Fifth Avenue, New York, New York 10110

thamesandhudsonusa.com

Library of Congress Catalog Card Number 2015932485

ISBN 978-0-500-34304-3

Printed and bound in China by Shanghai Offset Printing Products Limited

SuperLux

Smart Light Art, Design and Architecture for Cities

Editor

Davina Jackson

Essays

Mary-Anne Kyriakou
Vesna Petresin
Thomas Schielke
Peter Weibel

Comment

Peter Droege

Thames & Hudson

Contents

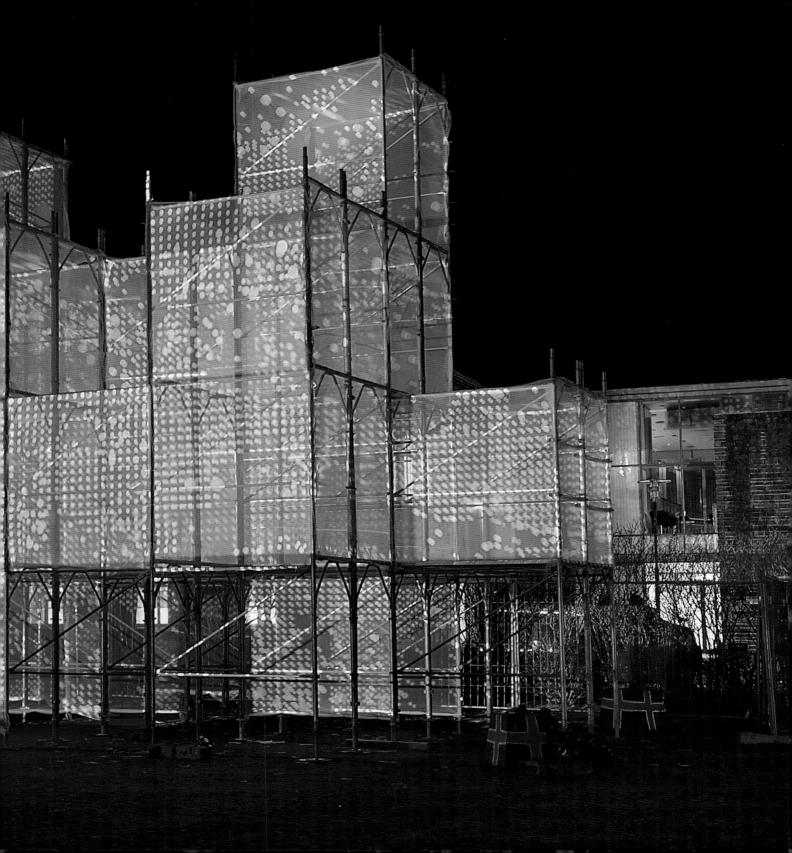

Cities of Cool Light

Introduction
Davina Jackson

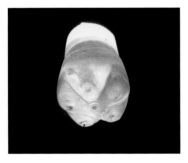

PET bottles, devised by the Brazilian inventor Alfredo Moser, make effective lights for simple shelters in developing countries.

Both gas and electric lights illuminated the Eiffel Tower when it first opened for the Paris Exposition of 1889 (photo US Library of Congress).

Today's smartest lighting innovation has no need for machine-generated power. Simply fill a clear plastic bottle with water (plus bleach) and silicon-seal it through a hole in the roof of any rudimentary shelter. Behold: daylight floods the dark interior.

PET bottles for conducting radiance, solar-powered LED lamps for night visibility and satellite-enabled smartphones to exchange instant knowledge globally: these are the 21st century's keys to illuminating billions of people living rough in settlements. Like the original campsites of London, New York, São Paulo and Sydney, some of these slums will become great global centres.

Light is the universal source of all human potential to act (even to eat) and achieve. After millennia of human reliance on flames, Thomas Edison demonstrated the first commercially viable electric lamp in 1887. He switched on the modern era of relentlessly radiant cities, beginning with the Eiffel Tower in 1889 and underpinning the utopian mid-20th-century visions of Le Corbusier, Mies van der Rohe with Richard Kelly, and other visionary urbanists on both sides of the Atlantic.

Flying over high-rise New York during the pre-1960 years of electricity-enabled commercial air travel, American atomic physicist Nick Holonyak, Jr. (then a young researcher for General Electric, the company Edison founded) told himself there must be a smarter way to light cities. He knew that incandescent lamps convert only 10% of their electrical supply to light and the other 90% is lost to heat.

'RED HOT' was the coverline of science journal *IEEE Spectrum* when it featured Holonyak's 1962 demonstration of the first visible red laser beam, which he generated from a pill-like semiconductor that was later labelled 'the magic one'. This white disc, incorporating tiny cleaved mirrors, was the world's first effective light-emitting diode (LED). It was made from its inventor's own alloy of gallium arsenide phosphide (abbreviated to GaAsP in science's periodic table of elements).

Holonyak's momentous creation of that first viable LED was followed by LEDs in yellow and green, before Japanese inventor Shuji Nakamura, working for Nichia Corporation, demonstrated the first high-intensity blue LED in 1994, and then proved that coating blue LEDs with a yellowish (YAG) phosphor could generate high-intensity white light. Nakamura and collaborators Isamu Akasaki and Hiroshi Amano won the Nobel Prize for Physics in 2014.

Holonyak's invention triggered the late 20th century's vast revolution in electronic equipment, including today's networked tablets and mobile phones. Nakamura not only enabled the current Blu-Ray DVD units and many ambitious RGB light and video spectacles in public venues, but also, crucially, created the LED non-colours (warm or cool white) that are essential for everyday lighting of contemporary spaces.

Holonyak and Nakamura consecutively pioneered today's global revolution in solid state lighting (SSL), where, in Holonyak's words, 'the current itself is the light'. After many years of refinement by major lighting manufacturers, LED lamps now use up to 75% less energy and last up to 25 times longer than incandescent or halogen bulbs, and up to three times longer than compact fluorescent light (CFL) tubes. Further efficiencies are possible when LED lights are powered with energy from renewable sources, including solar, wind and biofuel systems.

The new frontier for eco-efficient light in cities is the smart grid, which allows power to be shared among users on local neighbourhood networks – a strategy likely to minimize wastage of electricity over the longer distances necessary with metropolitan systems.

Another important goal is to reduce light pollution of the skies above cities. Is it a vain hope that parents might once again point their children to see stars from urban locations? The most promising strategy is to install new networks of dimmable LED streetlights and new fixtures designed to block uplighting and spread beams downwards.

We also need to find alternatives to carbon-belching fireworks and photon-spraying sky lasers for celebrating public occasions. Two internationally applauded events launched in Sydney – the 'Earth Hour' switch-off-lights demonstration in 2007 and the 'Smart Light Sydney' exhibition of non-carbon light artworks in 2009 – were international milestones highlighting new, ecologically responsible agendas among younger generations of cultural leaders.

Thanks to Holonyak and his followers over the past half-century, we now revel in a third age of light: the digitally networked LED era of cool-to-touch electroluminescence. Their advances in atom-related physics sparked unprecedented creative potentials for the internet-era 'City of Bits' heralded in 1995 by MIT architecture scholar, William J. Mitchell. Professor Mitchell also propagated the term 'smart cities', which inspired our choices here of 'smart light' projects.

This book is the world's first comprehensively illustrated monograph surveying contemporary landmarks and outstanding creativity in the emerging movement of smart (electroluminescent) light for urban art, architecture and environments. It contains more than 400 images of post-2008 examples of energy-effective light installations: illuminated buildings, bridges, streets, parks, plazas, media walls, public interiors, gallery spaces and water systems, including interactions and augmented reality games using mobile devices. Also included is an illustrated timeline of historical achievements with luminous structures, including paint, perforations, projections and pixel screens. Finally we compare night-lit maps of cities, using photographs taken by European astronauts (notably André Kuipers), optimized with the International Space Station's Nightpod (orbit speed focus correction) equipment.

Our essayists – Mary-Anne Kyriakou, Vesna Petresin, Thomas Schielke and Peter Weibel – provide uniquely informed perspectives on today's converging conditions. To conclude, urbanism expert Peter Droege challenges us, extravagant exploiters of light, to accelerate 100% reliance on renewable sources of power.

American physicist Nick Holonyak, Jr. demonstrated the world's first visible red LED lamp in 1962, using a pill-sized semiconductor.

Copenhagen has regulations to keep the night sky dark and clear, most notably with cautious lighting along the Stroget retail strip.

Overleaf: Lantern-lighting of ETFE air cushions cladding the Water Cube aquatic centre built for the 2008 Beijing Olympics (details p. 264).

Light's Bright Future: Lightscapes

Peter Weibel

'So what is light?' asked German physicist Heinrich Hertz in 1889. Hertz (after whom the scientific unit of wave frequency is named) worked on the cusp of humanity's transitions from understanding something of natural electricity to replicating various of its sequences in controlled conditions. His proof of the electromagnetic theory of waves allowed modern technology experts to begin to create and control the vast potentials of artificial light and light art.

Influenced by Michael Faraday (1791–1867) and the mathematical theories of James Clerk Maxwell (1831–1879), Hertz (1857–1894) concluded: 'Light is an electric phenomenon, light per se, all light, the light of the sun, the light of a candle, the light of a glow worm. Take the electricity out of the world and the light disappears.'

Today we think of both light and electricity as 'man-made' phenomena. To understand how they will continue to enable and express human progress (concentrated in cities), it is necessary to grasp how the history of artistic light (central in the broad evolution from artificial electricity to electronic arts) has developed.

Particle, Wave and Optics Theories

After the Renaissance, many physicists, such as Isaac Newton (1642–1727; *New Theory About Light and Colours*, 1672, and *Opticks*, 1704), Pierre-Simon Laplace (1749–1827) and David Brewster (1781–1868), believed that light consisted of the smallest of particles: corpuscles (tiny cells). This theory, proposing that particles were emitted in straight lines and at high speed by luminescent bodies, explained many phenomena of light – especially how white light refracted at the edges into its spectral colours.

Dutch scientist Christiaan Huygens (1629–1695) introduced the long-controversial wave theory of light in his 1690 *Traité de la lumière*. Not until the works of Thomas Young (1773–1829) and Augustin-Jean Fresnel (1788–1827), and Maxwell's 1865 *A Dynamical Theory of the Electromagnetic Field*, was Newton's corpuscle theory replaced by the wave theory, which Hertz proved and elaborated in 1886–88.

Wave theory more convincingly explains effects such as refraction and diffraction, which underpin modern knowledge of how light behaves. 'Since the days of Young and Fresnel we have known that light is a wave movement,' Hertz wrote. 'We know the speed of the waves, we know their length, we know that they are distortional waves.'

German physicist Heinrich Hertz, author of the electromagnetic theory of waves.

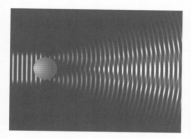

Diffraction is the process of light waves being diverted and expanded by an obstacle.

An orbital angular momentum (OAM) dish can spin vortex radio waves carrying parallel streams of massive data.

Refraction differs from diffraction in being a matter of particle optics. On the adjacent surfaces of two media – for example, air and clear plastic, or a water drop – refraction causes each ray of light to bend. The dependence of the refraction on wavelength is termed 'dispersion'. Each ray curves with a constant change in the dispersion speed. When we see refraction we witness light's past, as it were.

Diffraction refers to the waves being diverted by an obstacle in a medium. Thus a wave may spread in space if its direct route is blocked. Italian mathematician Francesco Maria Grimaldi (1618–1663) observed how sun rays that enter a room through a small opening create a larger patch of light than conventional optics would suggest, meaning the light rays expanded (diffracted). Diffractive optics describes the expansion of light no longer as rays but as waves in a pond. Certainly the future of light art will depend on diffractive optics.

Light art's future also relies on spiral dynamics for conveying data. Electromagnetic fields can shift into a helix mode, and some can spin either to the left or right. Since a demonstration in Venice in 2011 of 'vortex radio waves' by Swedish physicist Bo Thidé (b. 1948) and an Italian research team around Fabrizio Tamburini (b. 1963), it has been considered possible to use these 'spun' waves for radio transmissions that broadcast 'potentially infinite' data streams in parallel on the same frequency (Tamburini, 2011). Using several 'spin stages' (orbital angular momentum, or OAM, states), data transmission rates can be raised significantly. Light's OAM can thus be used to code and process information in multi-dimensional quantum spaces. Light therefore broadens from the optical dimension into a medium of information (data).

After Albert Einstein's (1879–1955) photon theory was published in 1905, light again was assigned corpuscular properties, which is why today we speak of light's wave–particle dualism. Both his photon theory of light and quantum theory enabled completely new models of light activity. If gravitation is used as the lens (which is the case with black holes), if photons are interlinked and the wave function of photons leads to a revolution in optic manipulation, and if light is neither a wave nor a corpuscle but a vortex (a specific type of light ray), then 'ghost images' can occur. Depending on the photon phases, the OAM spiral spectrum and the observer, these are comparable to a projected hologram without glass.

Ghost Cities

'Ghost imaging' is a technique that enables a high-resolution camera to produce images of an object that the camera itself does not see. This phenomenon is based on the quantum nature of light. Quantum correlations between pairs of photons that hit the object reach the camera lens on different paths. This difference can then be used to construct an image of the 'invisible' object. In 2009, a team of Chinese scientists, Xi-Hao Chen, Qian Liu, Kai-Hong Luo and Ling-An Wu, demonstrated ghost images and ghost diffraction using a single-pixel detector of 'true thermal' light. Photorefractive materials such as crystals, polymers and liquid crystal cells may improve the future viability of refractive optics: they could serve as optic-holographic data storage facilities.

In the age of molecular chemistry and protein chips, future laser and material technologies represent a controlled advance on phosphorescence and fluorescence (the spontaneous emission of light after the material has been agitated). Fluorescence polarization, fluorescence correlation spectroscopy and other procedures that exploit the fluorescent properties of fluorophores will enable new screens and displays, in new shapes and forms.

Ghost images can be 'accidentally' captured by high-resolution cameras when pairs of light photons bounce to the lens on different paths.

A giant LG curved OLED screen at London's Piccadilly Circus.

Today, light's most transformative technology involves luminescent diodes, light-emitting diodes (LEDs) or light diodes, the light-emitting semiconductor element. If electric current flows through the diodes, they radiate light, meaning a LED essentially entails electroluminescence.

Development of transistors in semiconductor physics after the early 1950s allowed the first visible spectrum LED lamps to be invented by American physicist Nick Holonyak, Jr. (b. 1928). Mass manufacture of LEDs is now allowing light art via movies and TV to become a mass-cultural phenomenon, transforming the world's commercial zones at night. OLEDs (organic light-emitting diodes) will change normal windows into TV screens and normal screens into tools of visibility. The walls of rooms, the façades of houses and the faces of cities will be transformed by OLEDs.

Building façades can already operate all hours as LED or OLED screens. Our cities are increasingly becoming mega-screens. Smartphones can currently be used to remotely control distant light sources, and are themselves becoming light sources that can be monitored by satellite cameras.

At concerts, mobile phones acting as electromagnetic candles can enable audiences to help control the music and the images on stage. Music performances are morphing into festivals of participation. Each person becomes a transmitter, a light artist. Not only signals such as text messages are being sent by electromagnetic waves, but also light messages. A second generation of ghost images is evolving.

Cities are changing from purely functional organizations to spectacles of colour, light, motion and sound. These are dream cities. No longer invented and operated as mere tourist attractions, they are the natural habitats of digital natives. They are sites and scenery for self-representation, moving poems of light, places where light and shadow shift like Gutenberg's moving letter blocks.

We act across cities as stages elaborated with texts, props, music and strange costumes. People promenade through cities as landscapes of longing, as virtual panoramas of imaginary journeys around the world, across invented continents oscillating between man-made nature and dreams. Every urban thing – every façade and every mobile object or subject – can be a screen. In this global context of *summa theatralica*, light is the vital medium of urbanity: cities are increasingly understood as 'lightscapes'.

Today's cities are fields of lambency. They are occupied by countless light sources, which increasingly are geo-tagged with their addresses (fixed or flowing co-ordinates) in space. These clusters will increasingly be manipulated creatively, as pixels on screens, by their originators, addressees, and near and far observers (both real local and virtually mediated remote audiences). Urban lightscapes will thus become individual light paintings: everyone will be able to sketch imagery across the light field of a city. Interactions of smartphones and OLED communications will generate city-wide ghost images. Cities are becoming polychromatic ghost towns of light.

Origins of Modern Electric Arts

What is light? In the fifty years after Hertz asked that question and proved the electric wave theory, Europe's most influential artists have disputed light's definition in visual terms.

'Colour is light,' said Dutch painter Vincent van Gogh (1854–1890).

'Light is colour,' said Hungarian painter and Bauhaus professor,
László Moholy-Nagy (1895–1946).

Until the end of the 19th century, painting mostly depicted sidereal light (from the sun and stars). After the electric light bulb evolved to carry commercial potential in the 1880s, innovative art practices shifted from representations of light to realities of light. Artists began to work with real light … not depicting illusions of natural light, but really using artificial light.

Just as Renaissance spring clocks had introduced a shift from sidereal time (derived from the stars) to artificial chronometry, so electricity initiated a switch from the natural light of the sun to the artificial light of the lamp. Light was no longer captured but exuded. The artwork became the generator, or emitter, of real light.

During this paradigm shift, colour took on an absolute status as the medium of the painting, beginning with 19th-century Impressionism and peaking with monochromatic panel paintings in the 20th century. The idea of primary colours, which dominated 20th-century abstract painting, was developed by scientists in the previous century. In 1802, Thomas Young had suggested that the eye perceives only red, green and violet, and that different mixtures of these produced all the other colours. This tri-chromatic theory was repeatedly demonstrated and today is considered scientifically proven. Among various technological advances, it has facilitated colour television, graphical user interfaces for computer screens and projections of images. The three cones of the human retina correspond to the on-screen colour dots of red, green and blue (RGB), which are positioned so close to one another that they mix because of the eye's limited capacity for resolution. The current exemplar of this principle is 'retina displays' on tablet devices, where pixels of colour are tiny and so plentiful that they fuse densely, like a felted fabric.

Baltic-German chemist Wilhelm Ostwald's theory of energetics, published in his *Vorlesungen über Naturphilosophie* of 1905, was especially influential on artists in the Constructivist, Bauhaus and de Stijl camps of the early 1920s. Ostwald (1853–1932) promoted four primary colours (including sea-green) in his 1916 colour system, but his key scientific claim was that our eyes see nothing but radiant energy, which triggers chemical changes on the cornea that we sense as light. With this concept, painters could use colour to portray light, and the prismatic development of colours played a decisive role even for the human perception of light.

Appreciating colour as a form of energy, as radiant energy, as electromagnetic waves, made it easier for artists to substitute light for paint/colour, as light is nothing other than energy and electromagnetic waves. In this way, colour became a phenomenon of light. Light became the overarching concept. It became an independent means, a medium and a material. This ignited the art of light.

Among contemporary painters, the 20th century is considered the era of the triumph of abstraction over figurative imagery. However, abstract painting did not affect the technical basis of panel paintings (canvas, stretcher frame, oil paint). The 20th century's crucial advance for art was, rather, a radical change and expansion of the technical medium on which an image could be based, namely converting panels of paint into screens displaying light. Conceptual debates between figurative and abstract should be transcended by concerns about a different pair of opposites: material and immaterial.

A replica of Bauhaus pioneer László Moholy-Nagy's 1920s 'Light-Space Modulator' (photo Bauhaus Dessau).

The sign of modernism is light. The 19th-century revolution in colour and paint, which triggered all the declarations of independence in modern art (the independence of colour, of surface and of form), led finally to the dethroning of colour, then paint (just another material). Immateriality and real light entered into art.

Notably, in 1930, László Moholy-Nagy developed a Light Prop, a kinetic 'space modulator' consisting of light, Plexiglas and steel. In a 1936 issue of the Hungarian journal *Telehor*, he outlined almost the entire spectrum of future light art with these notes:

1. the play of light outdoors space: a. light adverts, they still usually make use of linear surface effects [...]. b. then there are spotlight cannons (u.s.a. – companies and persil have done similar things), and c. as a future form the projection onto clouds, gaseous projection surfaces which you can walk through, for example d. as a future form the urban light theatre that one will experience from dirigibles, airplanes in the immense expanse of the light network, the shifts and changes of light surfaces, and which can definitely lead to a new, enhanced form of community festivals. 2. the play of light indoors: a. film with its new potential for projecting coloured, plastic and simultaneous effects: i.e., projected either with an enlarged number of apparatuses on a single point or onto all walls of the room, b. reflectory play of light, achieved with coloured projectors stencils or the like, such as laszlo's colour organ, either as standalone achievements or using tv and broadcast from a radio center, c. this also includes the colour piano, an instrument with a keyboard, a series of lamp units, and used to illuminate and highlight shadow screens. (that will be the future instrument of 'drawing instructor' for many demonstrations.) d. the light fresco, used to use artificial light to create large architectural units, buildings, wings, walls according to a fixed lighting plan. (it is highly probable that in future apartments will have a specific place reserved to receive these light frescoes, like a radio.)

German artist Nikolaus Braun (1900–1950), a member of the November Group, used actual electric light in his 'light reliefs' of the early 1920s. For him, light was material and, through light, colour/paint could be 'dematerialized'. Braun wrote:

The three basic factors in the visual arts – colour, form and light – have always contested with one another for pride of place. If we follow the individual lines of development in these basic factors then we will see an ever clearer perception of all three and a drive to give them concrete shape...

... I designed my sculptures and light reliefs to which I added the concrete property of electric light such that these structures were given a real light of their own.

This trend toward material dissolution is clearly expressed by a key figure in the use of new technical materials and light sources: Czech kinetic artist Zdeněk Pešánek (1896–1965), who in 1929–30 was probably the first artist to use neon. By 1925 Pešánek had created light projections using his spectrophone, a colour piano and luminodynamic sculptures. In 1930, he created the model for a light-kinetic sculpture (made of plaster, metal and a neon tube) for the Prague electricity utility, then made two similar fountains for the 1937 Paris world's fair, using colour lamps that could be programmed.

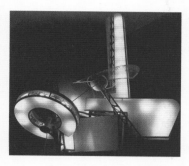

Czech artist Zdeněk Pešánek's late 1920s kinetic light sculpture for the Edison Power Station in Prague.

In the light experiments of the 1920s and 1930s, the transformation of material boxes into light boxes and of material reliefs into light reliefs (from Moholy-Nagy to Pešánek) formed the main basis for historical transitions in uses of light in art.

In the mid-1940s, the Arte Madi movement was founded in Buenos Aires. This group exhibited mobile kinetic sculptures, used unusual materials such as neon and fluorescent lamps, lasers, holographs and remote-controlled objects, and challenged viewers to participate in the transformable works.

In Italy in 1948, the first manifesto of *Spazialismo*, a group founded by Argentine-Italian artist Lucio Fontana (1899–1968), called for a radical inclusion of technology in forms of artistic expression: 'For this reason, we will make use of modern technological means to create artificial forms such as marvelous rainbows in the heavens. By means of radio and TV we will broadcast completely new forms of artistic expression.'

In 1949, Fontana built his first 'Ambiente spaziale' with spatial shapes and ultraviolet light, and perforated card and canvas to let light shine through. In 1952, the fifth *Manifesto spaziale* ('By contrast, we want to liberate art from its bondage to material') was broadcast on television, and TV art was thus inaugurated. Linking art to the development of new technical means, such as neon, TV and radar (all of which overcame the limits of time and place, and implemented the desire to liberate art from material), laid foundations for that cult of emptiness and light that dominated the artistic worlds of not only Fontana but also French artist Yves Klein (1928–1962) and Italian artist Piero Manzoni (1933–1963).

Dissolving Materiality

Around 1952, Yves Klein began painting monochrome pictures, and by 1958 his works were mainly painted with 'International Klein Blue'. That year, he realized his first 'immaterial' demonstration – an empty room in Galerie Iris Clert, Paris – to show 'the presence of pictorial sensibility in a state of *prima materia*'. Three years later he created his first fire pictures, using strong gas burners as 'brushes' applied to carton surfaces which dissolved into smoke.

The 1920s and 1930s' transformation of the image from colour technology to light technology peaked in the 1950s and 1960s with completely new forms of art, such as lumino-dynamic sculptures, light ballets, glass montages, light boxes and neon objects. Artists leading these changes included Nicolas Schöffer, Otto Piene, Adolf Luther, Hugo Demarco, Marc Adrian, Getulio Alviani, Martha Boto, the ZERO group, the Recherche d'Art Visuel (GRAV) group, Julio Le Parc, François Morellet and Dan Flavin, among others.

Diffusions and modulations of light led to light rooms and light environments by many modern artists, notably Gianni Colombo (1937–1993) and James Turrell (b. 1943). With Dan Flavin (1933–1996), light images expanded to constitute light spaces. In these works, and in pieces that used LEDs and light panels, by Bill Bell (b. 1928), Jenny Holzer (b. 1950) and others, artificial light supplanted the conceptual relevance of the panel picture format.

The immaterial itself finally became the material of painting. Around 1960, the ZERO movement (Günther Uecker, Heinz Mack, Otto Piene and allied artists such as Hermann Goepfert) made light the central theme and main medium of their art, with reflecting reliefs, metal elements in images and light machines.

In 1963, Flavin began working with fluorescent light (but not neon). While the light boxes had held light captive, here light was free and able to disseminate in space.

All four walls, the floor and the ceiling became the surface of the picture. Real light itself became the art.

The way toward dematerialization via white light was paved by the 'white manifestos' of Kazimir Malevich (1879–1935) through to Fontana, and the large white paintings by Robert Rauschenberg (1925–2008), Manzoni and others. With white paintings, pictorial field and environment tended to blend into an optical *Ganzfeld*, or 'overall field'. This concept encouraged Turrell's new expressions of time and space: in the 1980s, he called his light-filled experiential spaces 'Ganzfeld Pieces'.

Synchronies, Synchromies and Synaesthetics

Concordances, correspondences, parallelisms between eye and ear, between painting and music: for centuries these have been among the constants of experimental physiology and art. Technologies have always been used to stimulate different senses simultaneously – for example, the cosmogonic system devised by Athanasius Kircher (1601/2–1680) who created the first book on light art, entitled *Ars magna lucis et umbrae*, in 1671.

From 1877 to the First World War, various inventive musical artists (initially Bainbridge Bishop and Alexander Burnett Hector, and later Alexander Wallace Rimington and Preston S. Millar) demonstrated colour organs that could project coloured lights onto screens as effects of the notes being played. Nineteenth-century art's focus on absolute colour as the medium of light led to Rimington's 1912 book, *Colour-Music*. His celebrated colour organ projected many colours onto gauze-like screens and attracted interest from composers Richard Wagner and George Grove.

Russian painter Vladimir Baranoff-Rossine (1899–1944) used painted panes of glass for projections in his 1923 *Piano optophonique* performance. In 1925, French designer Paul Poiret (1879–1944) presented a light organ at one of his fashion shows, and Hungarian composer Alexander László (1895–1970) constructed a device, the 'sonchromatoscope', that projected coloured light effects synchronized to the musical score. In London in 1926, Adrian Bernard Leopold Klein (1892–1969) published *Colour-Music: The Art of Light* to promote the hypothesis that a key source for the origin of light art was the synaesthetic dreams of colour music and optophonics. In Graz, Austria, at the end of the 1920s, a Baltic musician, Anatol Vietinghoff-Scheel (1899–1933), experimented with the harmony of colour and music. His 'chromatophone' projected light from spots onto gauze-like screens along with colour films, accompanied by the music of Debussy or Scriabin. Charles Dockum (1904–1977) spent decades developing an appliance for projecting coloured light, finally presenting his 'Mobilcolor Projector' in 1952 at the Guggenheim Museum in New York.

The audio-visual world exploded with the genesis of the computer. The moving colour and light images that Thomas Wilfred (1889–1968) had, with great refinement, created by hand in the 1930s are now as ubiquitous as computer screensaver sequences, accessible at the press of a key.

American filmmaker Mary Ellen Bute (1906–1983) called her aesthetic credo 'seeing sound', which became the slogan of the music television industry. With custom-made machines, she used light as a drawing tool in the media of film and oscilloscopes. In the 1950s, mathematician Ben F. Laposky (1914–2000) also began working with oscilloscopes, producing waves shaped like analog curves. Using modified oscilloscopes, Laposky was able to shape, combine, modulate and transform the waves. He termed his oscilloscopic art 'a kind of visual music'.

At the end of the 1960s, video cameras enabled the first attempts at electronic imagery. These culminated in the 1970s and 1980s in computer-generated or supported images. Optophonetic coloured light films, coloured light music, light films: all made breakthroughs to transform mainstream culture, from the disco to the laser show.

Perfecting the Optophone

Dreams of 'chromophony' and 'optophony', of colour pianos and music painting, pervaded the 1920s artistic movements of Dada and the Bauhaus. In particular, Raoul Hausmann (1886–1971) became famous for his optophone, an invention he patented in 1926, whereby light rays were transposed electronically by means of selenium cells into sound waves, and sound waves were turned into light.

In Hausmann's 1933 manifesto, *Die überzüchteten Künste*, he wrote:

Gentlemen Musicians and Gentlemen Painters, you will see through your ears and hear through your eyes! The electrical spectrophone destroys all notions of sound, colour and shape.

The optophone or spectrophone, a kind of colour piano, functions with a keyboard not unlike a calculator's, with about 100 keys and 100 fields with different surface textures made of chrome gelatin, whose spectral shifts in line are transferred by the ray of a neon lamp onto a photocell or optic collector. The resulting changes in coloured shapes are projected onto a screen, while the photocells transform the light values into electrical charges that appear in the loudspeakers as acoustic effects. On an optophone, optophonetic compositions are played.

Hausmann's visions are now exemplified by Kurt Laurenz Theinert's performances with the 'Visual Piano', a digital audio-visual instrument with a MIDI keyboard, modified to add six pedals and using custom open source software programmed by Roland Blach (v1.0) and Philipp Rahlenbeck (v2.0). The keyboard, pedals and trackpad play notes and synchronized graphic elements that are projected 360° around an indoor or outdoor chamber (see p.90).

Today's most ubiquitous optophone, the mobile smart device, is the actual technical highpoint of the dreamed analogies of sound and image. Although bereft of any artistic concept and without any artistic correspondence, smartphones enable the production, distribution and review of images and sounds. They give us a (tele)voice, a (tele)vision and remote listening, in a single medium, thus fulfilling the dream of colour music, the unity of sound and vision.

Today, the entire development of light art, past and future, can be studied against the horizon of this fresh technological format, a mass-communication device that can simultaneously stimulate the senses and record most varieties of human creative expression.

Expect a torrent of new art via apps. Apps are the next destiny, and commodity, of the optophone's trajectory in the history of art.

Smartphones: today's answer to Raoul Hausmann's late 1920s visions for optophones.

Elevations

Opposite: Lantern lighting of the world's largest tent structure, the Khan Shatyr Entertainment Centre, in Astana, Kazakhstan, by Norman Foster with Claude Engle and Sembol (details p. 264).

Architectural Light

Architecture's vital poetic purpose is to sculpt light. Throughout history, artists have needed light – from the sky and from flames, then from cameras and electric lamps – to create atmospheres that satisfy human emotions and to convince audiences to see their structures as works of art.

Local vagaries of natural light are essential factors in the *genius loci*, spirit of place, which inspires architects to draw site-specific visions. In 1973, influential American architect Louis Kahn told an audience at New York's Pratt Institute: 'Light is really the source of all being.'

European modern architecture's greatest thought leader, Le Corbusier, indoctrinated generations of students with this 1923 concept:

Architecture is the masterly, correct and magnificent play of masses brought together in light. Our eyes are made to see forms in light; light and shade reveal these forms...

Corb was born in 1887, the year that Thomas Edison demonstrated the first effective electric incandescent lamp. Both men understood that artificial light would enable all the new architectural technologies that were emerging from the industrial revolution. Since the publication of his main manifestos in the 1920s, Corb's proposals for the *Ville Radieuse* (Radiant City) of gleaming towers have influenced planning of new urban areas throughout the world.

Paris switched on the world's first electrically floodlit buildings in 1889. The Eiffel Tower and the Place de l'Opéra were lit up to promote the 'City of Light' and to advertise that year's May–October Paris Exposition.

From the 1890s to the First World War, Chicago was the world centre for building high stacks of offices and lighting them externally with powerful arc lamps; it was followed by New York with its much taller skyscrapers between the First and Second World Wars. In both cities, theatre and film experts were asked to arrange arc lamps to highlight the sculptural qualities of the new commercial icons – but their spectacular effects began to be criticized as too vulgar for permanent public installations.

In an interview for General Electric's 1930 *Architecture of the Night* booklet, New York architect Raymond Hood suggested that beyond floodlighting, 'colour, pattern and even "movement" may be attempted'. After his experiments with Johann (John) Kliegel (a founder of the

Kliegel Brothers Universal Electric Stage Lighting Company) to test lighting strategies for his American Radiator Building, Hood said:

We tried multi-colored revolving lights and produced at one time the effect of the building's being on fire. We threw spots of light on jets of steam rising out of the smoke-stack. Then again, with moving lights, we had the whole top of the building waving like a tree in a strong wind. With cross-lighting, that is to say, lighting from different sources and different directions across the same forms, the most unusual cubistic patterns were developed.

Hood advised that these experiments should not be installed permanently because 'at present the art is new, our knowledge very scant and we all play safe'. Instead he encouraged vertical lighting from below, because it:

... adds to the element of mystery, as the fading out of the lights from the bottom to the top exaggerates the perspective, and seeing the building disappearing up into the night gives it an increased height. ... the lights can be arranged to stream up the vertical forms ... and the setbacks and terraces provide ideal places for the operation of the lights.

In another article for the *Architecture of the Night* booklet, GE's director of illumination engineering, Walter D'Arcy Ryan, contrarily suggested that the floodlighting of tall structures should not merely fade out, but instead could be 'surmounted by a coloured, or much more intensely illuminated, element ... in the form of a spire, tower, lantern or dome'.

As early 20th-century glass manufacturers gradually increased the sizes of window panels, architects were able to design entire buildings like lanterns, or shoji screens, where interior lights would create the radiance. Dutch and German modernists led the renaissance of this strategy in Europe – especially with factories, where brick structures darkly framed horizontal arrays of backlit windows.

The German term *lichtarchitektur* was introduced in 1927 by theorist Joachim Teichmüller, who was involved in avant-garde debates (initiated by architect Bruno Taut's pre-1920s concepts) about the implications of 'crystalline' architecture. Teichmüller and other influential writers emphasized a need for co-operation and transfers of specialist information from various relevant disciplines such as theatre design, engineering, architecture, art and science.

Should there be today a specific discipline of professionals officially called light architects? Authorities controlling registration of architects are strongly resistant to expanding uses of this term because they prefer to expand the professional activities of registered architects. Also expanding is multi-disciplinary

interest in creative ways to illuminate buildings ... suggesting that light architects would not be as tightly focused a group as, say, landscape architects.

Global lighting manufacturers, such as Philips, General Electric, ERCO, Zumtobel and Sylvania, now employ sophisticated design leaders to consult on creative solutions for high-profile projects implementing their products. Alongside this, major international structural engineering companies are allocating high salaries to young electrical engineers with expertise in planning digital multimedia installations that will dynamically activate static structures at night.

Lighting manufacturers began their commercial rollout of LED products for architectural uses only in the early 2000s. By then, many creative potentials of all three RGB colours (together producing white) had been tested with experimental performances at art and design exhibitions, trade shows, theatres and concerts, public festivals and globally televised sports events such as the Olympics. Kinetic art shows and prizes, either sponsored by lighting suppliers or organized by specialist centres including ZKM in Karlsruhe, Germany, and Ars Electronica in Linz, Austria, are also catalysts for pioneering demonstrations of architectural lighting.

Computer simulations of dynamic lighting schemes, especially daylight effects on buildings, are becoming integrated with architectural design software to pre-test concepts 'in the virtual model'. American computer scientist Greg Ward is credited for inventing the first lighting simulation program, Radiance, in 1985. Released for use by other programmers in 2002, its simulation engine now underpins the lighting design components of other widely used CAD (computer-aided design) and BIM (building information modelling) packages.

The classical approaches to lighting buildings for outdoor viewing remain floodlighting (often with various kinds of spotlights) and backlighting (where lights inside a building glow through its transparent areas). But the technology revolution triggered by semiconductors (for LEDs specifically and computers generally) has vastly expanded these potentials to include 3D projection mapping (the latest and most sophisticated technique for building floodlighting) as well as various kinds of media façades, ranging from giant OLED (organic LED) television screens that curve around corners to low- and high-resolution 'pixel displays', where large arrays of LEDs are fixed to buildings and individually controlled via automated computer programs. These dynamic systems are igniting and mutating architecture's ancient conceptual basis of stasis.

Seafoam Filigree

Mer-veille (MuCEM)
AIK (Yann Kersalé), Rudy Ricciotti, 8'18"

Marseille, France

Pulses of ultramarine and turquoise light ripple like zephyrs across the steel lace façades of Marseille's new Museum of European and Mediterranean Civilizations (MuCEM). Designed by light artist Yann Kersalé with architect Rudy Ricciotti and technical light designers 8'18", the *Mer-veille* art concept incorporates references to sea waves lit by fishermen's torches and to ancient Moorish architectural lantern effects with carved filigree screens. Signalling like a lighthouse to observers on both land and water, the new MuCEM 'building within a building' includes a 72 sq m (775 ft²) outer structure of ultra high performance fibre concrete (UHPFC) lace screens and a 52 sq m (560 ft²) inner structure, containing the extensively glazed galleries and facilities. It was built on the former J4 docklands site near the city's historic Fort Saint-Jean.

Technology | 200 x LEC Lyon 4240-Havre customized spotlights (façades), 112 x LEC Lyon Belval 5631 LED luminaires (ramp and moats), 30 x LEC Lyon Bordeaux 5630 LED luminaires (roof terrace).

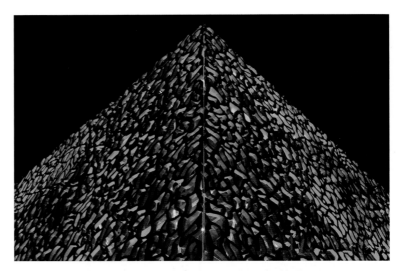

Elevations

Architectural Light

Monocoque Marvel

Yas Island Marina Hotel
Asymptote Architecture, Arup

Abu Dhabi, UAE

Abu Dhabi's tradition of net fishing is referenced in the gravity-defying design of two monocoques (shells of stainless steel and glass) hovering above the Yas Island Marina Hotel. The hotel's architecture, by Asymptote with Arup, comprises two ten-storey, elliptical buildings arranged in a T-format (one above the marina waterline and one on land). These are connected by a fuselage-styled bridge and brise-soleil across stacked loops of Abu Dhabi's new Formula 1 racetrack. The complex is distinguished by a steel and glass diagrid exoskeleton (serpentine shell), which includes real-time sensor-mediated light and ventilation systems, and comes alive with custom-programmed LED colour sequences at night. The engineers balanced a desire for the skin to reflect sky and surroundings while avoiding sun glare for race-car drivers and pilots of low-flying aircraft.

Technology | 5,850+ Enfis bespoke 4-channel UNO Plus RGBW LED arrays, e:cue LAS 5.1 control drivers, DMX RDM temperature monitoring, LED strips integrated below concrete path kerbs, Edison Price ceramic metal halide 3000K PAR 30 exterior downlights.

Tubular Gymnastics

Melbourne Theatre Company
Electrolight

Melbourne, Australia

Metal halide lamps crisply illuminate white pipework to create a web-like illusion around the 'black box' structure of a new venue for the Melbourne Theatre Company in Australia. This building, housing two theatre spaces, was inserted near the eastern entry of the Crown Towers hotel in the Southbank arts precinct, with its flytower (needed for back-stage scenery-changing equipment) projecting from the original building. Lighting specialists Electrolight worked with ARM Architecture to highlight the pipework using Sill 030 70W metal halide ceramic fixtures, fixed at a 10° beam angle to avoid light spilling onto the dark building walls. Each night these are turned on via a photocell and off with a timer.

Technology | 23 x Sill 030 linear projector lights with 70W 4000K metal halide ceramic lamps, DMX control system.

Arboreal Visuals

Wintergarden
Studio 505, Xenian, Ramus Illumination

Brisbane, Australia

Spanning almost a city block across three faces of the Wintergarden Shopping Centre is a metal screen of lacy layers, emulating a forest pulsing with luminous imagery. Incorporating 24,000 LEDs playing low-resolution video imagery at night, the *Wintergarden*'s 90m (295 ft) edifice is made of layered steel sheets laser-cut in outlines inspired by foliage in winter. Mall owner ISPT and developer Brookfield Multiplex created this magnetic civic and retail branding display with façade architects Studio 505, specialist engineer Tensys, lighting effects designer Ramus Illumination, consulting artist John Warwicker and contractors Urban Art Projects. The lighting system, installed by Xenian, uses Philips's Color Kinetics iColor Flex LMX flexible strands of fifty individually addressable LED nodes, linked to a Coolux Pandoras Box media server control system. As well as playing 180 hours of video imagery created by concert effects expert Bruce Ramus, the mall's managers can curate the content to include local artists and community groups and to promote seasonal events and themes.

Technology | Philips Color Kinetics iColor Flex LMX nodes, Coolux Pandoras Box media server.

032

Circular Logic

Silo 468
Lighting Design Collective

Helsinki, Finland

After nightfall on a weather-whipped clifftop near Helsinki, a drum-shaped white pavilion appears to pulsate with sparks of light. Its rhythms are driven by software downloading the latest weather data and by algorithms derived from movements of swarms. To help activate a new 'district of light' for 11,000 future citizens, this obsolete oil silo (Silo 468) has been transformed into a light artwork and community hall. Lighting Design Collective, led by Tapio Rosenius, worked with the City of Helsinki, Helsinki Energy and TASKE on the project. The steel cylinder, measuring 36m (118 ft) in diameter and 17m (nearly 56 ft) in height, was perforated with 2,012 holes (symbolizing Helsinki's year as the World Design Capital). A 128 x 10 grid of 1,280 2700K warm white LEDs and 450 pivoting steel mirrors was fixed around the structure's red-painted inside wall, shining light through the holes.

Technology | 1,280 x 2700K warm white LEDs, 450 pivoting steel mirrors, bespoke software coded with OpenFrameworks, e:cue Lighting Control Engine mx server.

034

Lustrous Structure

Hatoya-3 Building
Forlights, Nōsu

Tokyo, Japan

Mysteriously lit in a tartan pattern, the Hatoya-3 apartment building occupies a busy corner of the old town in Japan's capital, Tokyo. Designed by architects Nōsu and lighting designer Forlights (Yutaka Inaba), the building is externally distinguished by its red, white and green pattern of vertical panels inspired by the neighbourhood's traditional advertising signs. Instead of Japanese calligraphy sales messages, these 'signboards' form geometrically elegant Piet Mondrian-style façades comprising three materials: red paint, frosted glass and ivy vines (green trellises to screen balconies). Defining vertical edges of the panels are aluminium spandrels, projected beyond adjacent surfaces to create distinctive shadow effects. At night, the façades are illuminated by various lengths of Morikawa LED/REVOX R-CX 150 multi-LED lamp consoles, fixed to 'graze' light at low angles across the 'signboards'. These lamps cause the flickering shadows and candle-lantern effects that remind observers of Japan's elegant cultural history before electricity sparked the nightlife of its contemporary cities.

Technology | 102 x Morikawa LED/REVOX R-CX 150 outdoor multi-LED consoles of varying lengths (total 491 x lamps).

Elevations

Architectural Light

Colour Perspectives

Fetch
Erwin Redl

Columbus, USA

This page | After dark, lines of multi-coloured light rotate within the white steel scaffolding that distinguishes the 158m-long (518 ft) east façade of the Wexner Center for the Arts in Columbus, Ohio. New York light artist Erwin Redl created this installation in 2010 to activate the internationally celebrated Wexner complex designed by architect Peter Eisenman in 1989 ... and to pay homage to legendary art precedents by painter Sol LeWitt and pioneer human animation photographer Eadweard Muybridge. Redl produced his animation illusion by digitally pulsing light in rapid sequences along three rows of incrementally twisted RGB LED tubes. Each tube was centred within each 'box' along the 20m-high (65 ft) scaffold structure, which was illuminated with different colours to create light perspective effects.

Technology | 100 x bespoke 122cm-long (48 in) RGB LED tubes, Nicolaude ESA Pro DMX controller, Meanwell power supplies.

State Theatre Centre of Western Australia
Electrolight, Kerry Hill Architects

Perth, Australia

Opposite page | Western Australia's new centre for theatre performances, designed by Kerry Hill Architects, incorporates dramatic spaces, materials, colours and illumination. The lighting scheme, by Electrolight, includes splendid surprises and trompe l'œil effects to appropriately 'paint with light' every area of the building. Five basic strategies underpinned the light planning: wall washing, lighting people, reflection, integration with the architecture, and care in using luminaires with different characteristics and colour temperatures. WE-EF metal halide luminaires were selected especially to accentuate the Bronze Box, a towering and dense screen of gilded bronze pipes highlighting the grand foyer staircase.

Technology | Flos (from Euroluce) Antares pinhole 35W IRC halogen recessed downlights, WE-EF (from Eagle Lighting) FLC141s 150W, 250W and 350W 3000K metal halide lamps, ERCO track and 50W halogen wall washers, Philips Dynalite control system.

Elevations

Architectural Light

Amazing Games

Gran Casino Costa Brava
artec3 Studio, b720 Fermín Vázquez Arquitectos

Girona, Spain

With its two main façades sparkling towards night traffic on Girona's busy Lloret de Mar intersection, the Gran Casino Costa Brava is an elongated ensemble of charcoal-toned, irregularly angled forms. Tucked at the base of a hill below the existing Hotel Monterrey, the casino now extends the hotel's garden across its roof. The inclined glass planes of the casino's two street-facing façades are illuminated with a networked matrix of 700mm-long (27½ in) RGB colour-changing LED tubes, designed for this project by lighting consultants artec3 Studio. Half of the luminaires are fixed horizontally at 1 metre (3 ft 3 in) centres between twin columns of cables inclined to match the glazed walls; the remainder are suspended vertically as pendant lamps.

Technology | Matrix of 469 x horizontal and 462 x pendant SMD RGB LED tubes (each 40mm x 700mm; 1½ in x 27½ in) with 9 x LED 3W 24V DC IP65 controls every 30 luminaires, 3 DMX channels.

Architectural Light

Sky Lines

MetaLicht
Mischa Kuball

Wuppertal, Germany

Tower blocks on the main Grifflenberg campus of the
Bergische Universität Wuppertal have been activated
with softly brightening/fading lines of white LED lighting
(totalling 900 metres, or over 2,950 ft). Cologne light artist
and professor Mischa Kuball created this nightly installation
for the 40th anniversary of the university in 2012. Titled
MetaLicht, it was conceived to dynamically signal creative
and intellectually uplifting connections between the
university and the community of Wuppertal and the
surrounding Bergisches Land.

Technology | 900m (approx. 2,950 ft) Zumtobel Tecton 6000K white LED tubes,
DMX control system, three wind power generators.

Elevations

Architectural Light

History Alive

Blackburn Town Hall
Studio Fink

Blackburn, UK

Thirty-six unobtrusive colour-changing RGB LED lamps have been installed in three arrays to illuminate key features of the classical stone edifice of Britain's 1852 Blackburn Town Hall. Mostly casting white light (colours on special occasions), the fixtures are drilled at mortar joints to reduce permanent damage to the heritage structure.

Technology | 36 x Philips Color Kinetics ColorBlast RGB LED lamps.

Donaustrom
Waltraut Cooper

Vienna, Austria

Eight Corinthian columns support the pediment-entablature of Austria's 1883 Greek Revival Parliament House, dominating Vienna's Ringstraße. In 2007, 2008 and 2010, light artist Waltraut Cooper temporarily transformed these monumental columns into 'virtual waterfalls' using 9m-long (29½ ft) channels of blue LED light, slotted into a central flute of each column. *Donaustrom* means 'the flow of the Danube river', but 'strom' also suggests electric current – a point relevant to present-day hydro-electric power stations along the Danube.

Technology | 8 x 9m-long (29½ ft) blue LED luminaires, custom-made by KLW using aluminium U channels with blue acrylic diffuser panels, with pre-programmed light animation.

Speicherstadt
Michael Batz

Hamburg, Germany

Hamburg's Speicherstadt, a dense district of late 19th-century red-brick warehouses lining narrow canals, is part of the major Hafen-City urban renewal project. Hamburg light artist Michael Batz illuminated this cluster of warehouse apartments with a romantic, chiaroscuro lighting concept, using over 1,000 Philips conventional and LED light fixtures, averaging less than 25 watts energy use per unit.

Technology | 1,000 x Philips MVF 606 Decoflood metal halide lamps, 40 x Philips 2 60° spots, 200 x Norca 60W tubes.

Reichstag
Michael Batz

Berlin, Germany

Light artist Michael Batz won a 2008 competition with three
scenarios to relight Germany's federal Parliament building,
the Reichstag in Berlin. His main scheme bathes all four
façades, including the roof sculptures and outdoor stairs,
in warm white light from 400 LED spots and metal halide
vapour lamps.

Technology | 200 x Philips DecoScene DBP521 CDM-TM 35W/NB metal halide
lamps, 200 x Philips Decoflood DCP623/626 CDM-TM 35W lamps, 16 x Philips
LEDline 2 BCS716 lamps, 28 x Philips LEDline2 BCS710 lamps.

Desert Spires

Capital Gate
DPA Lighting, RMJM

Abu Dhabi, UAE

This page | Inclining 18° westwards at its peak of 160 metres (525 ft), the Capital Gate tower is spectacularly illuminated at night via a double-glazed diagrid framing system incorporating 686 Traxon Wall Washer Shield XB colour-changing RGB LED luminaires fitted at junctures around the external lattice. All lighting is orchestrated by an electronic control system which can uniquely instruct each lamp in the network. Developed as a conference, exhibition, office and hotel centre, the complex includes a large array of solar panels for electricity and hot water efficiency.

Technology | 686 x Traxon Wall Washer Shield XB RGB LED luminaires, electronic control system.

Burj Khalifa Tower
Fisher Marantz Stone, Speirs + Major, Skidmore, Owings & Merrill

Dubai, UAE

Opposite page | Burj Khalifa, the world's tallest building at 828 metres (approximately 2,720 ft) and 160 floors, is elegantly illuminated externally with metal halide uplights, metal halide floodlights and halogen downlights, fixed to balconies, parapets and setback terraces winding around the tapered spire. All lighting and data fittings are connected to a Philips Dynalite modular control system, installed with technical consultants Tectronics and principal lighting designers Fisher Marantz Stone. London-based lighting designers Speirs + Major also provided a 'celebratory layer' using high-powered stroboscopes.

Technology | Philips Dynalite building lighting system including 7,000+ modular controllers and 14,000+ Philips Revolution Series 2 user-operation panels; 'Celebration Layer' including 868 x Dataflash stroboscopes, 6 x Fineline searchlights, ETC Congo light servers, ETC Unison Paradigm control system, programmed with Capture visualization software.

Elevations

Architectural Light

City Screens

Blatant transmitters of fleeting incidents and eternal desires, electronic billboards pulse packets of light across hectic urban transport streams. Catching the latent sentiments of humans in transit, these screens of flickering visuals are today's signs of any city's intensity of integration with global flows of commerce.

Times Square, Seoul Square, Piccadilly, Shibuya and Orchard Road: these are our planet's most spectacular crossroads of night trading and constant traffic. All these junctions are memorable for giant televisions activating the podiums, portals and pinnacles of their perimeter buildings.

Rather than broadcasting 'programs', as with living room televisions, they are washing viewers' brains with optically arousing pageants of video clips; they are channelling alternative versions of desire.

Last century's Marxists could see these political conduits coming to us citizens of their future. Here is a prediction from *The Society of the Spectacle* (1967, clauses 24, 34) by Guy Debord, a legendary instigator of the Paris student riots of May 1968:

The spectacle is the ruling order's non-stop discourse about itself, its never-ending monologue of self-praise, its self-portrait at the stage of totalitarian domination of all aspects of life. ... the spectacle is capital accumulated to the point that it becomes images.

Contemporary media façades and urban screens follow the first electric advertising sign, spelling 'EDISON' to visitors at the London electricity exposition of 1882. A decade later, New York's first electric sign used 1,457 incandescent bulbs to promote 'Manhattan Beach – Swept by Ocean Breezes' to viewers near Broadway, part of the zone named Times Square in 1904.

Co-branded with New York's legendary newspaper, Times Square has a history that is a reliable chronology and case study of international advances and convergences in urban lighting, street advertising, entertainment publicity and multimedia communications. Even before the First World War, the area's jumble of electric signs, with dramatic floodlighting of Manhattan's nearby early skyscrapers, won New York its indelible reputation as the world's greatest metropolis of the modern electric 20th century.

No rival city emerged until the mid-1980s, when Japan celebrated its then-new economic supremacy over American manufacturers of consumer products. Its first hurrah was a Sony JumboTron screen (25 x 40m; 82 x 130 ft) at the Tokyo Expo in 1985, followed in 1986 by the first outdoor television overlooking the scramble crossing beside Shibuya's JR train station.

Urban screens today are a serious focus for digital media artists and scholars. Academic crucibles include the MIT Media Lab and *Leonardo* electronic arts journal in Cambridge, Massachusetts; ISEA International symposia on the electronic arts and SIGGRAPH conferences on computer graphics; the festivals and prizes presented each September by the Ars Electronica Center in Linz; academic journals such as Sage's *Society and Culture*; and Europe's new Media Architecture Institute and Institute of Network Cultures. Creative professionals demonstrate their prowess via special events, concert tours, city light festivals and permanent screens installed on prominent commercial buildings.

Like earlier neon billboards and home televisions, video screens are electrically powered, high-density arrays of fine lines of radiant light. The imagery appears via binary (on/off) computer code: a lightning-fast, visual form of the manual dot/dash message signals developed by American telegraph engineer Samuel Morse in the late 1830s.

Video displays (the latest being curved LED or OLED panels) are not the only format of urban screens. Less sophisticated visually, but more adaptable to irregular building forms, are pixel screens, comprising 'picture elements' or points of light. First these used incandescent bulbs, as at Times Square in the early 20th century, then various forms of fluorescent or neon tubes and bulbs, such as the compact fluoro lamps arrayed across the animistic curves of the Kunsthaus at Graz in 2003, or, more recently, the 2,200 sq m (24,000 ft²) GreenPix Zero Energy Media Wall of 2,292 solar-powered RGB LEDs installed on the Xicui Centre in Beijing in 2008.

Lighting engineers are frustrated by many of the current technical barriers to realizing their dream of presenting all kinds of images at any time, seamlessly integrated with complex structural forms. Sydney-based light artist M. Hank Haeusler has published the most ambitious list of desirable future capabilities for media façades. He demands environmentally ethical sources of energy, minimal requirements for data and power cables and connections, individually addressable and removable pixels to allow openings to be designed within each screen, and robust construction allowing easy maintenance and replaceable components.

Haeusler and others have been designing pixel screens in irregular 3D formats. They seek maximum adjustability of aspect ratios, pixel resolutions and screen sizes, screen viewing angles up to 360°, and their choice of any form of voxels (3D pixels: the first voxel screen was ETH-Z's NOVA system, see p.238).

Leaders of the international Media Architecture Institute, and their research and industry empathizers, aim to accelerate the capabilities of media façades so that modern architecture's 'curtain walls' (literally, drops of glass from the edges of building floor plates) will be superseded by elaborately wrapped 'textiles' of light.

This vision updates the stained glass spectacles of medieval cathedrals and adds moving images to German sociologist Gottfried Semper's mid-19th-century Platonic ideal of architecture originating as wood-framed huts, draped with woven mats as their rudimentary 'walls'.

Indeed, traditional tribal rug patterns influenced UK/US architect Christopher Alexander's post-1960s mathematical concepts for programming environments in both cities and computers. He and fellow 1960s-'70s Londoners Cedric Price and Gordon Pask are admired by today's media architects for inventing many concepts that now inform both the software structures and visual content of multimedia environments.

What's next with public screens? Researchers at Germany's Max Planck Institute for Informatics, led by Martin Fuchs (now of Stuttgart University), have tested 6D hologram displays, where images can respond to changes of both viewing point and light. Fuchs and others seem to be on the verge of achieving techniques that could convert city monuments into giant ghostly holograms, pulsing to twilight transitions from sunlight to darkness – and to movements by outdoor audiences. That style of cinema-meets-architecture imagineering would have delighted Walt Disney ... and astonished his Renaissance predecessors inventing the earliest magic lanterns.

At their most subversive, today's media architects want to replace the modernist definition of architecture, as compositions of solids and voids 'under light', with a new dream (perhaps fallacious but certainly courageous) of architecture *as* light.

Box of Treasures

Cleusa Gift Store
Studio ix

São Paulo, Brazil

Conjuring visions of a mystical box of treasures, the four-storey Cleusa gift store has revitalized a prominent corner of São Paulo's prestigious Jardins shopping district. After a major renovation, new curtain walls of fritted glass glow with delicate colour tints cast by computer-controlled RGB LED strips. Commissioned by the Santa Helena Group, architects Jayme Mestieri and Carlos Azevedo demolished the masonry façades and replaced the interiors of an existing commercial building, then worked with Studio ix light designer Guinter Parschalk on imagineering new 'light box' effects. Glass columns and four-armed steel brackets now support the new glass façades 60cm (23½ in) beyond the edges of the building's floor slabs, where the 59 RGB LED bars have been installed. The glass has been fritted with two layers of translucently screenprinted film, providing both decoration (dots forming a large rose motif) and sun filtering, while delivering spectacles for shoppers.

Technology | 59 x LED bars with ellipsoidal optics, MR-16 LED spots and T5 tubes in coves, integrated with a DMX control network.

Elevations

City Screens

Market Organics

ION Orchard
Benoy Associates, Arup

Singapore

Aerial waves of glass and steel pulsate with rippling light sparks, dazzling drivers and wanderers across Singapore's most hectic intersection of the Orchard Road strip. This is the geometrically awesome, LED-studded diagrid roofshell, or monocoque, that shelters and signals the eight-storey ION Orchard retail and civic centre, above an MRT station. British architects Benoy Associates and engineers Arup designed the multi-curved podium roof and two media/light-equipped façade systems for development consortium Orchard Turn. The design alludes to natural foliage forms in tropical jungles. On 'skin' areas of the façade and canopy, LED luminaires are mounted on plates approximately 1.5m (5 ft) apart, allowing low-resolution routines of colour-changing lights, while on 'seed' areas the LEDs are mounted on aluminium frames (carrying cables) that fit within the diagrid's modules. The main media wall, measuring 16 x 20m (52½ x 65 ft), sits tightly behind double-glazing in a prominent zone of the street façade, delivering high-resolution TV broadcasts with minimal interference from reflections. The entire roofshell is supported by multi-branched steel columns designed with increasingly fine, tapered diameters, like the natural growth patterns of fast-growing trees.

Technology | Krislite Krisledz Mahina Node RGB LED luminaires for roofshell, Tiara Linear and Lumina Vision panels for media screens, with control systems.

Elevations

City Screens

052

Licht Lanterns

LICHT
Götz Lemberg

Stuttgart, Germany

Opposite page | In a variation of *Weiss* (see below), Berlin-based light, sound and space artist Götz Lemberg created an illuminated display across five glassy floors of the Stuttgart headquarters for the State Bank of Baden-Württemberg (LBBW). He formed the logo 'LICHT' with capital letters 2.5m high (over 8 ft) per floor, using 20mm (¾ in) neon tubes in red, orange, yellow, green and blue. With grandMA software linked to a grandMA lighting controller, he dissolved the logo into dynamic lines and different colours and words over a 15-minute performance loop. The installation was a highlight of the 2010 'Musikfest Stuttgart'.

Technology | 20mm (¾ in) coloured neon tubes, controller, programming.

Weiss
Götz Lemberg

Stuttgart, Germany

This page | One word, 'LICHT', boldly dominates a high floor of the glass-fronted art museum in Stuttgart. Observers wandering around the plaza are intrigued to watch subtle dissolves and fluxes of the illuminated, geometrically multi-coloured letters. These seem to shatter into fragments of colour, the shards multiplied as glass-veiled reflections, optically smudged by the atmosphere of the city at night. Götz Lemberg developed this temporary installation to 'toy with the meanings and imagery of a word and the perception of forms and patterns'.

Technology | Red, green and blue squares of light, mirrored blocks, reflective foil, controller, dimmer.

Video Urbanism

Linien
Ursula Scherrer

Seoul, Korea

This page | South Korea's capital boasts one of the world's largest urban screens: a 99 x 78m (325 x 256 ft) 'media canvas' with 42,000 LED pixels, occupying the entire façade of the Gana Art building, on the square outside Seoul's main railway station. Gana curates and constantly broadcasts international video art on its façade, a strategy that promotes its rising art-curatorial status via a torrent of YouTube videos uploaded by the artists. One optically arresting video installation was *Linien* (Lines) by Swiss-born, New York-based artist Ursula Scherrer, broadcast during Gana's 'Media Arts from Switzerland' festival in 2012–13. While preparing the video, Scherrer wrote a concept poem highlighting her idea of dancing, digital, lineal patterns: 'lines that create order, order that falls apart' ... 'lines/twisting around each other/parallel/that never meet/creating a space/then disappear'. The work draws on her experience as a choreographer, photographer, mixed media artist, poet and theorist, and her impulses to fuse physical and virtual creativity.

Technology | Galaxia Electronics 99 x 78m (325 x 256 ft) (332 x 264 to 996 x 792 pixels) media canvas with 42,000 individually addressable 5.5mm (¼ in) LED lamps, controlled by up to two computers, single-channel video content.

El Molino
artec3 Studio, BOPBAA

Barcelona, Spain

Opposite page | One of Barcelona's historic nightlife magnets, El Molino (The Mill), a burlesque theatre and cocktail lounge, has been rejuvenated with LED illuminations of its façade. Architects BOPBAA and lighting experts artec3 Studio installed a concave screen of 39 x 237 LED pixels at the top of the four-storey building, fitting 237 SMD RGB colour-changing lamps behind perforated translucent diffusers along 39 vertical aluminium strips fixed 39–40cm (approx. 15½ in) apart. The designers also fixed LED strip lighting around the theatre's iconic sign of four rotating windmill blades (installed in the 1940s as a homage to the Moulin Rouge cabaret in Paris).

Technology | Philips Color Kinetics iColorFlex SLX RGB IP66 LED strips with DMX control system, translucent PMMA diffusers, control system, bespoke low-res video content.

Elevations

City Screens

Helio Stature

Miroir de Mer (Sea Mirror)
Yann Kersalé, Jean Nouvel

Sydney, Australia

Astonishing cantilevers (structural projections that appear poised for disastrous collapse) are historically significant marvels of late 20th-century modernist architecture. Today's most precarious example is a LED-studded aerial heliostat by Paris artist Yann Kersalé, extending west from the 29th storey of a Sydney apartment tower designed by Paris architect Jean Nouvel. Heliostats are mechanical (now also sensor- and computer-controlled) arrays of pivoting mirrors that can track and bounce the sun's rays (technically termed 'flux') to specific places that need more daylight. At Sydney's Central Park precinct, developed by Singapore-based Fraser Properties, Kersalé's 900 sq m (9,700 ft²) heliostat transmits daylight to dark zones of the site and provides a dazzling light art display at night. Fitted to the eastern tower, its 320 mirrors bounce sunrays from 40 larger motorized and mirrored panels on the roof of the western tower, down to the ground-level retail area, swimming pool and park. Kersalé, a yachtsman, designed four night lighting sequences, using the Philips Color Kinetics LED system, inspired by seasonal changes of sky-reflected colours across 'mythical' Sydney Harbour. Operating over four hours on four nights a week, these displays exploit 2,880 RGB LEDs, producing coloured lights that dynamically mix with distorted sky and city reflections in the mirrors. He describes the night presence of the heliostat as a 'geopoetical signal ... an allegory ... a symbol of the sea in the city'.

Technology | 2,880 x Philips Color Kinetics LED nodes, bespoke heliostat of 320 x 1.5 sq m (16 ft²) reflector panels (9 LEDs per panel) suspended from a 110 tonne cantilevered steel framework.

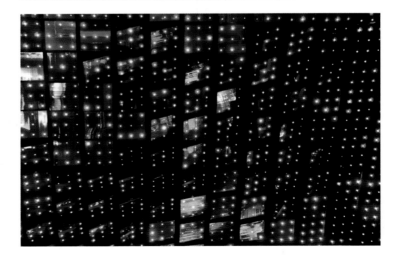

Elevations

City Screens

Word Plays

YOUR TEXT HERE
UNSTABLE (Marcos Zotes)

Detroit, USA

Opposite page | Submit text messages from your device to a website, then watch them appear enlarged on nearby buildings. This frontier between physical structures and virtual communications is forming a new creative genre known as 'augmented architecture'. Spanish-born, Iceland-based architect-artist Marcos Zotes, leading the UNSTABLE studio in Reykjavik, has produced several *YOUR TEXT HERE* projection performances to foster 'citizen proclamations' that highlight last century's domination of cities by commercial advertising signs. Zotes designed a screen interface with Scrabble-like tiles that could be relettered with texts sent to his project website. In 2012, he transformed the façade of a Detroit production plant, then buildings in New York and Reykjavik.

Technology | 1 x 20K lumens projector, Mac Syphon framework apps including MadMapper, Modul8.

Touch. Do Not Please The Work of Art
Cornelia Erdmann, Michael Lee Hong Hwee

Singapore and Sydney, Australia

This page | *Touch. Do Not Please the Work of Art* is a 1.7m-high (5½ ft), 40m-long (130 ft) lightwall, titled as an ironic comment on the paradox of 'touching' light, and social restrictions on how and when touching is acceptable. Designed by German light artist Cornelia Erdmann with Singapore artist Michael Lee Hong Hwee, it was installed first for Singapore's 2010 'iLight Marina Bay' festival, then Erdmann adapted it at Sydney's 2011 'Vivid' festival. Each wall is finished with paint and overlaid with glow-in-the-dark adhesive lettering, colour-matched so the message is not easily detected during the day. At night the walls are light-activated with LED or halogen floodlights that are timer-flashed every ten seconds for two seconds per pulse. Passers-by can cast fleeting traces of their silhouettes as 'shadow graffiti'.

Technology | Singapore: 8 x Philips 300W LED floodlights, timer. Sydney: 6 x 1000W halogen floodlights, timer.

Liquid Luz

Roca Barcelona Gallery
artec3 Studio

Barcelona, Spain

Beams of white light mimic rain, ripples and other fluid dimensions within the double-glass walls of the Roca bathroomware showroom on a Barcelona street corner. Designed by artec3 Studio with Office Architecture in Barcelona (OAB), this kinetic façade system uses IP67-rated, software-controlled LED tubes fitted along top, centre and bottom brackets in the wall cavity. The architectural vision involved transparency and elongation in the building's form (designed for the same city where Mies van der Rohe prototyped the original modern glass pavilion in 1929), with a surface treatment like 'liquid veils'. Fluidly dynamic effects are programmed from Visionarte software, controlling all luminaires (in groups of four) via four 512-channel universes of a DMX network control system. The glass box frames a double-height exhibition space with a dark concrete floor. The wall lighting filters and softens sunlight to allow viewing of sanitary fixtures in an atmosphere of 'intimacy [and] hygiene ... with an almost hypnotic mood'. The interior extends to conventionally walled office and service zones behind the exhibition space.

Technology | Troll ROC 6500K linear white LED spot projectors, Visionarte DMX control software (four universes).

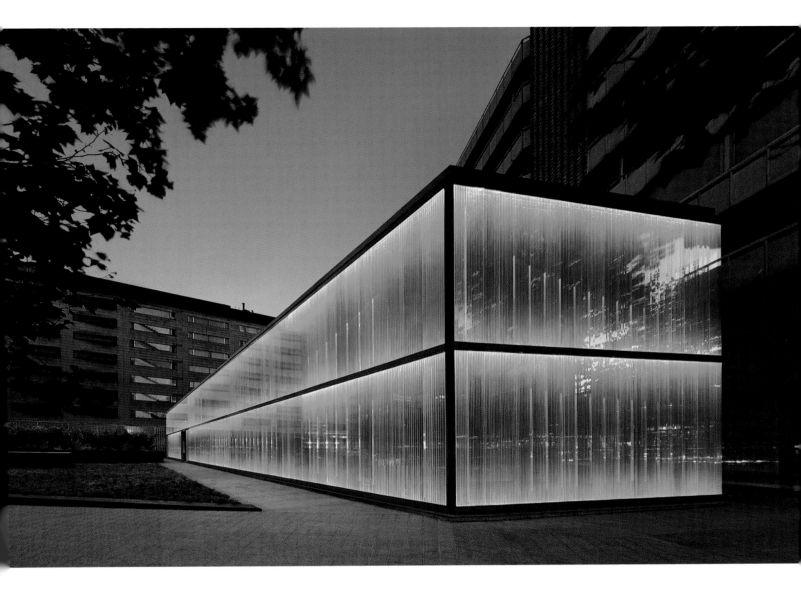

Elevations

City Screens

062

Civic Transmissions

Code (Version 2)
Laurens Tan

Beijing, China

Opposite page | Pervasive eyes blink, apparently randomly, at pedestrians wandering the Sanlitun commercial zone in Beijing's regenerating district of Chaoyang. Artist Laurens Tan produced this animated loop to simulate the computer coding used to learn words and sounds in the Putonghua (Mandarin) language. Titled *Code (Version 2)*, it was broadcast on the 24m-wide (80 ft) screen façade of the MegaBox cinema, promoting its 2009 festival of Australian films.

Technology | Single-channel video animation on infinite loop, 24m (80 ft) screen.

After Light
Storybox

Singapore

This page | Shipping containers are capsules of mystique. Whether stacked regimentally, floating lost at sea or cast aside near warehouses, these mundane metal crates transmit echoes of perilous ocean voyages, carrying precious cargoes. Storybox's *After Light* work at Singapore's 2012 'iLight Marina Bay' festival used 27 containers and 14 projection screens to broadcast visual explorations of different meanings of light. Three themes – divine light, industrial light and human light – were creatively interpreted as audio-video loops by the New Zealand-based Storybox team, led by Rob Appierdo. This work also included physical art installations by guest artists within four of the on-ground crates. *After Light* has been performed at waterfront light festivals in Petone and Auckland (New Zealand) and at Sydney's 'Vivid' festival.

Technology | 14 x Sony 2500 projectors, 8 x Apple Mac Mini computers, open source creative programming, 150m (490 ft) cables, 6 x 500W speakers.

Spectral Insights

Sprechende Wand (Talking Wall)
Daniel Hausig, with HBKsaar students

Backnang and Saarbrücken, Germany

This page top and bottom | With students from the Hochschule der Bildenden Künste Saar, German light artist Daniel Hausig installed RGB LEDs inside the historic town museum in Backnang and the HBKsaar building in Saarbrücken, then activated multi-coloured lantern effects with a custom-programmed lighting control system.

Technology | Colour-changing RGB LEDs, custom-programmed control system.

The Kind Spirits
Aleksandra Stratimirović

Ljubljana, Slovenia

Opposite top and bottom left | White spirits, all sharing a gentle and benevolent demeanour, shine mysteriously from the dark windows of a derelict building near the Shoemakers' Bridge in Slovenia's capital city, Ljubljana. A charming spectacle for citizens, *The Kind Spirits* was an art installation from Stockholm-based artist Aleksandra Stratimirović for Ljubljana's midsummer 2010 'Lighting Guerrilla' festival.

Technology | 30 x LED E4 lamps, 30 x transparent plastic shopping bags, nylon threads.

Floating Waters
Jens Schader

Wiesbaden, Germany

Opposite bottom right | Marine creatures seem to float darkly in tropically bright seawater, sharing the unlikely 'fishtank' of three top floors of Wiesbaden's Zircon Tower. German light artist Jens Schader, of the Raumbasis studio in Darmstadt, created this temporary 'kinetic shadow play' spectacle during a promotion of the tower's conversion to offices. For each of four months through early 2013, the black silhouettes of 'creatures' were seen suspended in different colour schemes of 'water'.

Technology | 84 x 15W PAR 56 RGB LED floodlights, 10 x 4W electric motors.

Elevations

City Screens

Corner Theatrics

2–22 La Vitrine culturelle
Moment Factory

Montreal, Canada

Digital creativity and transparency are key missions for
La Vitrine culturelle, the events-tickets centre marking the
St-Catherine and St-Laurent corner of Quartier des Spectacles
in Montreal. These themes are promoted via light art signals
and media façade treatments created by Moment Factory
with architects Ædifica and Gilles Huot, and Société de
Développement Angus. Glass-exposed passages around
three floors of the building's edges have been converted
into multimedia corridors lined with LED panels, forming
an 'urban theatre'. The panels have been carefully designed
to integrate with and highlight the existing architecture,
and they constantly show videos controlled via Moment
Factory's X-Agora system. External lighting of the building
includes narrow beam RGB LEDs angled to graze horizontally
across the façade, and diffused contour lights to delineate
the roofline. Atop La Vitrine's main corner, a 2-22 sign is
illuminated with white LEDs. The centre's design supports
the Quartier des Spectacles' vision for a 'Luminous Pathway'
linking its attractions.

Technology | 100 sq m (1,075 ft²) custom LED tiles with 18mm (¾ in) pixel pitch
and 6000 NITS brightness (for passages), 30m (100 ft) RGB LED grazing fixtures,
65m (215 ft) custom RGB LED exterior contour lights, custom RGB LED modules
(each approx 3.5 x 1.75m; 11 x 5½ ft), X-Agora (custom) multimedia control and
playback system.

Elevations

City Screens

Projecting Fantasies

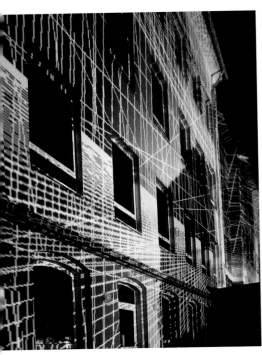

How does a historic stone monument suddenly crumble to rubble, without an explosion of dynamite? Then how can it be impeccably reconstructed just a few seconds later? Welcome to the awesome illusions being conjured by video artists in the magical new genre of architectural projections.

Also known as 3D projection mapping and spatial augmented reality, this globally popular trans-millennial technology mashes entertainment industry updates of photographic arts with traditional painting, sculpture and architectural drafting.

Urban projection artists first work in their design programs to apply astonishing video imagery onto geo-tagged virtual models of one or more city buildings. Then they go out at night to project the pre-mapped videos exactly onto the façades of their real structures.

The models are made by overlapping laser scans taken of each building by specialized cameras fixed to tripods, drones or light aeroplanes. The scans are digitally stitched together as 'point clouds' of data imagery. Each pixel of a cloud image is geo-tagged with the xyz co-ordinates of its corresponding point on the real building.

On site, one or several high-powered video projectors are pointed towards each building edifice, as with any projection screen. Final display details for the 3D-mapped video are rehearsed with a laptop computer linked to each projector. Often the visual narratives and special effects are accompanied by music and sound effects.

Today's urban projections often combine sophisticated video effects and motion graphics concepts from games, animations, cinema SFX, and image, movie and audio editing software. Artists are needed to personally create the multimedia works being shown, while scientists and electrical engineers are constantly refining, redesigning and expanding the technical capabilities of camera and projector equipment.

Finally, the projectionists offer their own stamps of creativity not only in editing the loops of imagery but also in their styles of performance – from the urban guerrilla (light bombing) tactics of electronic media students and digital street artists to formal orchestra-accompanied and acrobat-augmented outdoor cinema performances.

Contemporary projections can be mapped precisely onto geometrically complex structures, such as ornately carved classical buildings or large sculptures of animals and other organic forms. One outstanding example was London musical artist Brian Eno's *77 Million Paintings* display of imperceptibly morphing imagery, digitally mapped across the irregular curves of the Sydney Opera House for the first 'Vivid/Smart Light' festival in 2009 (see p.71).

Eno's Sydney projection and others recently shown as giant telegenic backdrops for Olympics, Expos and other global events are vastly more sophisticated than history's earliest projections of devils and demons via magic lanterns. These metal machines, fitted with rudimentary hand-ground lenses, were first described in the mid-16th century by Italian engineer Giovanni Fontana and were prototyped from the mid-17th century by Athanasius Kircher, Christiaan Huygens and others.

In 1895, the first projection of film (in tape rather than single slide format) was demonstrated in the French town of La Ciotat by brother inventors Auguste and Louis Lumière. They used (and later patented and improved) a 'Cinématographe' device invented by another French cinema pioneer, Léon Bouly.

The film projection genre has a non-projection precedent, with the first large 360° scenic panorama painted of Edinburgh by local artist Robert Barker in 1787. This and similar 360° murals of cities (each painted from a centrally located hill) were shown publicly around the inside walls of 'panorama rotundas' constructed in London's Leicester Square and European cities during the late 18th and early 19th centuries.

After the Lumières launched the first film projection, painted panoramas evolved to become projected photographic scenes known as 'cycloramas'. Further developments of panoramas, using photographic, cinema and eventually digital imaging systems, have been transferred to outdoor architectural projections.

The first modern spatially augmented projections were the spectral effects created by Disney imagineers Rolly Crump and Yale Gracey for the 'Haunted Mansion' attraction which opened at the California Disneyland in 1969 (after Walt Disney's death). Spectators in remote-controlled 'doom buggies' rode through the dark house of horrors, sighting occasional disembodied heads which appeared alive through projected animations. Today these horror projections might be done with .gif digital motion clips – today's update of the multiple slides used to show motion via magic lanterns and the 'flicks' of Disney's first cartoon animations.

In the late 1970s, just before Apple launched the personal computer, multimedia artist Michael Naimark – then a founding researcher at the MIT Media Lab – prototyped the Moving Movie and Talking Head techniques, which updated the Disney 3D spectral effects using a Super8 movie camera. He later invented other significant camera and projection techniques, including Spacecode for sensor-equipped cameras (1986) and Moviemaps (from 1986, and 3D versions from 1993). In 1998, the world's first workshop promoting 'spatial augmented reality' (SAR) was held in San Francisco by three young University of North Carolina researchers pioneering this approach: Ramesh Raskar, Greg Welch and Henry Fuchs. Their rationale for new 'computational illumination' techniques was:

Traditionally a projector is considered a display device that maps a 2D image onto a larger flat screen. We have questioned this restrictive consideration. By treating the projector as the dual of a camera – as a 3D projective device that matches 2D pixels onto 4D rays – we have shown that programmable illumination can be used in a versatile fashion for display, augmentation, optical communication and mobility.

Raskar, now an associate professor leading the MIT Media Lab's Camera Culture group, first studio-tested 3D architectural projections with a scale model of the (white) Taj Mahal surrounded by small projectors. He and his North Carolina colleagues called their system 'shader lamps' because their key concern was how to use multiple projectors to eliminate 'self-shadows' on the 'object'.

Potentials of architectural projections increased exponentially after Shuji Nakamura invented the first blue and blue-enabled white LED lamps in the mid-1990s, completing the RGB (red, green, blue = white) spectrum for mixing all colours visible via artificial light.

Another frontier for projection mapping has emerged with interactive light manipulation demonstrations at some city night festivals since the late 2000s. Users of software apps on laptops and mobile phones can now change the images and effects that are projection-mapped onto urban objects (one example was the popular *Light of the Merlion* demonstration by Portuguese artists Ocubo at Singapore's 2012 'iLight Marina Bay' festival, where excited children tapped laptop keys to 'paint-by-numbers' a patchwork design of light colours projected onto the city's famous colonial mascot sculpture; see p.206).

For young people considering creative careers, urban interaction design seems a major opportunity zone. There is scope to transform architectural projections from a 'broadcast' phenomenon (one experience to many viewers) into novel transmedia genres where viewers become players flowing between virtual and real-world domains of behaviour.

Radiant Mirages

77 Million Paintings
Brian Eno/Lumen London,
The Electric Canvas, Ramus Illumination

Sydney, Australia

British musician Brian Eno is a pioneer of ambient soundscapes. In recent years he has also been evolving ambient videoscapes that are broadcast either indoors on screens or projected across outdoor sites. His most iconic version of *77 Million Paintings* was 3D projection-mapped across (for the first time) both the east and west sides of the Sydney Opera House roofshells. It was performed over three weeks to highlight his *Luminous* series of events for 'Smart Light Sydney', a cornerstone of the inaugural 'Vivid' festival in 2009. Eno's Lumen London team worked with Sydney projection-mapping experts The Electric Canvas and Melbourne multimedia events specialist Bruce Ramus. To offset the high energy consumption of outdoor projections, the equipment was powered with locally sourced renewable biofuel.

Technology | Christie Digital 20K projectors, PIGI film-strip projectors, biofuelled electricity generators, OnlyView media servers, OnlyCue control system.

Terrain Scapes

Uhllich(t)
Ingo Bracke

Mosel Valley, Germany

This page | Beside Germany's River Mosel (connecting the ancient Roman Empire city of Trier with the Rhine) lies the Winninger Uhlen vineyard, with rows of vines streaking across the terraced lower slopes of surrounding mountains. In 2010, architect and light artist Ingo Bracke updated the post-1960s fabric-shrouding precedents of Christo and Jeanne-Claude by translucently transforming this pastoral landscape with layers of architectural diagrams. These images of overlaid cross-sections and site plans represented a palimpsest of visions for past and future impositions on the 'natural' environment. They were projected via customized glass slides (gobos) inside eight projectors.

Technology | 8 x bespoke high-efficiency gobo projectors with 575W MSR lamps.

Felsenzauber (Rock Magic)
Ingo Bracke

Bavarian Alps, Germany

Opposite page | Alluding to Christ's Stations of the Cross journey through Jerusalem, Ingo Bracke's *Felsenzauber* art ceremony involved a long walk punctuated by multimedia installations, performances, sounds of commotion and profound pauses. Participants rambled along a 3km (1 mile) mountain path including notable views and encounters with a watermill, forest, ruins, electricity plant, valley, dam, reservoir, waterfall and alpine meadow, all elaborated with sound and lightscapes. Different stories were expressed with Bracke's *Felsenzauber* performances in 2010 and 2014: light and the power of water (*Aus Wasser werde Licht*, celebrating the centenary of the power plant) and the mystery of Nature by night (*Felsenzauber für Nachtwandler*).

Technology | Gobo projectors (10 x MSR 575W, 2 x MSD 1200W/700W, 10 x 1000W halogen, 10 x 50W halogen) with various optics, 4 x Pani projectors (2000W and 5000W with hand-painted coloured slides), 5 x DLP video projectors using film renderings, 40 x 12x3W RGB LED lights, 10 x sound-interactive LED tubes, 40 x 1000W PAR 64 lights with gels, 4 x 250W Water Effect projectors, DMX programming, 10 x 3D sound systems.

Elevations   Projecting Fantasies

Creation Sensations

Ode à La Vie (Ode to Life)
Moment Factory

Barcelona, Spain

Opposite page | Barcelona's fantastic Sagrada Familia basilica was the 'canvas' for *Ode à La Vie*, a 2012 multimedia projection show celebrating the creation of the universe and the complexities of human life. The church's Nativity façade, approximately 150m (490 ft) tall and 30m (100 ft) wide, was transformed over three nights of 15-minute light art performances mapped onto the world's most sculpturally complex structure. These included imagery inspired by lively colour sketches drawn by the Sagrada Familia's key architect, Antoni Gaudí. Presented by the City of Montreal with a creative team led by Moment Factory, *Ode à La Vie* was conceived as 'a living fresco of colour, light and sound'. Comprised of seven acts, the performance required 16 video projectors and 56 moving lights, linked to 13 computers using Moment Factory's X-Agora playback control system.

Technology | 16 x Christie Digital 20K and 18K lumens video projectors, 56 x moving, 16 x strobe and 32 x LED PAR lights, X-Agora 3D-mapping software on 13 computers, 8 x smoke machines, 1,800 x LED butterflies per show.

Gutenberg
Casa Magica

Pittsburgh, USA

This page | Pittsburgh's iconic Cathedral of Learning – a classic American 1920s skyscraper in late Gothic Revival style – was transformed for the city's 2008 Festival of Lights. German projection artists Casa Magica (Friedrich Förster and Sabine Weissinger) shrouded the 163m-high (535 ft) tower with hot type lettering to honour Johannes Gutenberg's mid-15th-century invention of mechanical movable type printing (the foundation of modern literacy and mass education). As well as 2D imagery shrouding the tower's ornately decorated and stepped façades, the artists included 3D shadow motifs alluding to Gutenberg's block letters cast from metal (also acknowledging Pittsburgh's history as America's most productive steel manufacturing city).

Technology | 5 x Pani BP 12 Platinum HMI slide projectors, 2 x Pani BP 6 GT HMI projectors, 2 x Pani BP 4 GT slide projectors, Adobe Photoshop software.

Landmark Language

Text Spectacles
Detlef Hartung, Georg Trenz

Various locations

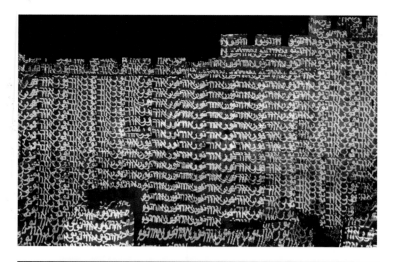

German light artists Detlef Hartung and Georg Trenz use graphically bold projections of words, in locally meaningful languages, to incite hope and cultural pride among their spectators.

This page top | *Light*, Jerusalem, Israel: Combinations of the English, Arabic and Hebrew letters used to form the primal word 'LIGHT' were projected as radiant, fluid patterns across Jerusalem's grittily eroded stone walls.

This page bottom | *Schauspiel* (Spectacle), Fürstenfeld, Germany: In a two-channel video projection performance for the 2011 'Brucker Kulturnacht', emotion-igniting words were fleetingly inscribed with rippling capital letters across the dense foliage of trees and hedges in the cultivated garden of Fürstenfeld Abbey on Bavaria's Amper floodplain.

Opposite top | *Wortschatz – Lebenszeichen* (Vocabulary – Sign of Life), St Goar, Germany: Gentle flows of selected German words, all set in one bold compressed typeface, were projected onto the Rhine Valley's symbolic Lorelei rock, and reflected in the gently flowing darkness of the river.

Opposite bottom | *Lebenszeichen* (Sign of Life), Koblenz, Germany: All sides of the Kaiser Wilhelm I monument in Koblenz were transformed with projected flows of nationally symbolic words, reprising neglected sources of German pride.

Technology | Location-specific combinations of Bright Sign HD 220 synchronized media players, Leica slide projectors, Pani BP 6 projectors with AMD32 slide changers and CS70 movie scrollers, Flying Pig Wholehog II lighting control system, Dataton WATCHOUT software and splitters.

Elevations   Projecting Fantasies

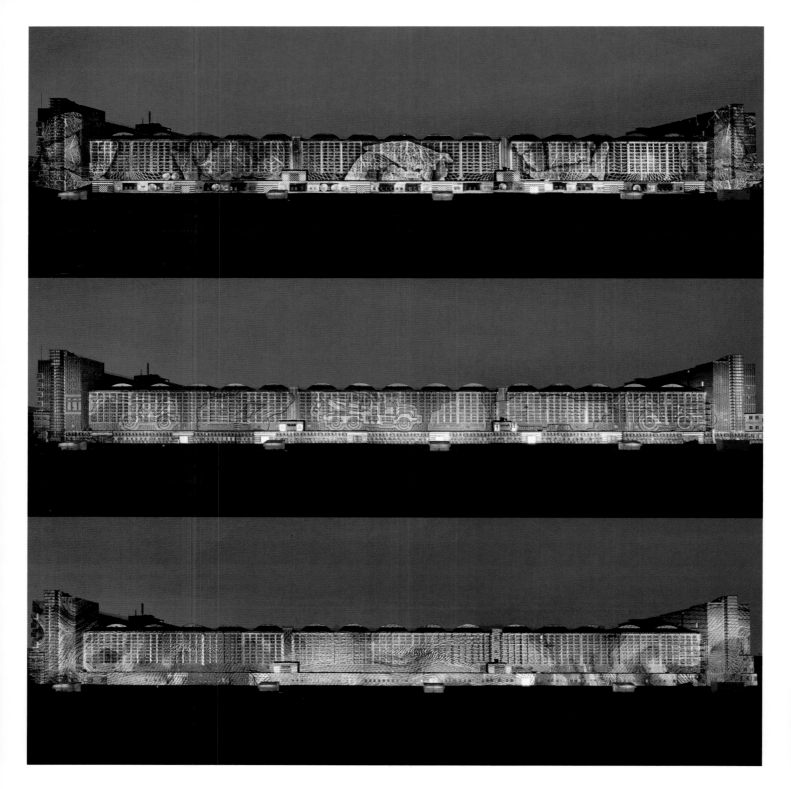

078

Scroll Scape

Main Embankment Panorama
Casa Magica

Frankfurt, Germany

Informal motifs in pastel tints glide across the 250m-long (820 ft) south façade of an abandoned 1920s brick warehouse on the north bank of Frankfurt's Main river. German art team Casa Magica (Friedrich Förster and Sabine Weissinger) used slide projections to depict the past, present and future of the building. Originally occupied by Frankfurt's produce markets, it was recently converted into headquarters for the European Central Bank. Commissioned by the bank to highlight its next premises as part of Frankfurt's 2008 'Luminale' festival, Casa Magica developed a cinematic 'night panorama'. They used eleven Pani HMI slide projectors (using Osram medium arc metal halide gas discharge bulbs) to display their customized images highlighting various uses of the building, including banknotes alluding to its continuity as a venue for exchanges of money.

Technology | 2 x Pani BP 12 Platinum HMI projectors, 9 x Pani BP 6 GT HMI projectors, Adobe Photoshop software, Stumpfl Wings Platinum control system.

Monument Movies

Night Beacon
Ian de Gruchy

Frankston, Australia

On Australia's sub-Antarctic coast near Melbourne, the beach suburb of Frankston hosts a midwinter light and multimedia arts festival. For the first show in 2012, projection artist Ian de Gruchy reinvented the town's Peninsula building (isolated and prominent on the waterfront) with a dissolve sequence of thirty images that lent new virtual architectural styles to the banal physical structure. The images were perspective-corrected in Photoshop and projected on two façades using two long-range slide projectors.

Technology | 2 x Pani BP 6 projectors, slides mapped in Photoshop.

Dr Who
Spinifex Group, Technical Direction Company

Sydney, Australia

Sydney's Customs House, a colonial heritage monument at the centre of Circular Quay, became the ornate canvas for an ambitious 2013 video projection to celebrate the 50th anniversary of *Dr Who*, one of the BBC's most successful television series. The 6-minute performance was created by James Tufrey of the Spinifex Group to include the Time Clock, TARDIS and other classic icons, as well as Venetian scenes promoting the then-forthcoming *Dr Who* Series 7. Hardware and video-mapping onto 3D laser scans of the building were provided by the Technical Direction Company.

Technology | 14 x Barco 26K projectors, Dataton WATCHOUT media servers.

City Life
The Electric Canvas

Sydney, Australia

Fantasy concepts of 'a city within a city' transformed the historic façade of Sydney's Customs House for the 2012 'Vivid' festival. The Electric Canvas created dynamic sequences using switches of perspective, converting the building's columns to streets and the central clock to a fountain, and animating the imaginary metropolis with bustling traffic, pedestrians, birds and other animated creatures, all to a custom-composed soundtrack.

Technology | Christie Digital video projectors, OnlyView media servers.

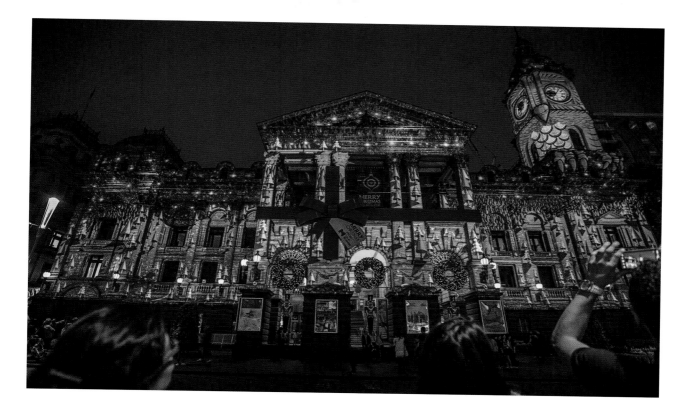

The Nights Before Christmas
The Electric Canvas

Melbourne, Australia

Witty fairyland animations played across the ornate stone façade
of Melbourne Town Hall, celebrating *The Nights Before Christmas*
in 2012. This video production converted the corner clocktower to
a jovial owl (with the clockfaces as its eyes) and reinterpreted the
façade as a fantastic toy workshop, with cartoon creatures scurrying
along its ledges and fascias. The imagery was mapped to the building
using architectural templates created from 3D modelling and POV
techniques, and accompanied by a soundtrack peppered with familiar
seasonal memories (also composed by The Electric Canvas).

Technology | Christie Digital video projectors, OnlyView media servers.

082

Single Vision

Awakening of Mary
Titia Ex

Amsterdam, The Netherlands, and Brussels, Belgium

Opposite page | Video close-ups of the faces of several contemporary women were morphed to create a new 3D televisual 'light sculpture' which ethereally reinterprets the Virgin Mary at a giant scale. This *Awakening of Mary* artwork was first projected on the façade of Amsterdam's Old Church in 2012, then on the Notre Dame de la Chapelle church in Brussels in 2013. Dutch artist Titia Ex aimed to spark a profound sense of intimacy, soulfulness and respect for women. Her spectacles were accompanied by the Nederlands Kamerkoor recording of Monteverdi's 1610 *Marian Vespers*.

Technology | Video loop comprising 8-9 clips of each model, 4K camera, 1 x Barco 18K projector, Coolux media server, projection cloth (netting).

CCTV/Creative Control
UNSTABLE

New York, USA

This page top | Beaming down to Brooklyn citizens is the Orwellian image of an all-seeing eyeball, graphically signifying the overview effects of today's urban paradigm of ubiquitous public surveillance cameras remotely controlled by police forces. This imagery, by artist Marcos Zotes of the UNSTABLE studio in Reykjavik, Iceland, was temporarily projected upwards to the underside of an abandoned water tower.

Technology | 1 x 6K lumens projector, MadMapper, Modul8 software.

Rafmögnuð Náttúra (Electric Nature)
UNSTABLE

Reykjavik, Iceland

This page centre and bottom | Competition-selected as the opening act for the 2012 'Winter Lights' festival in Reykjavik, *Rafmögnud Náttúra* was a spectacular, music-accompanied, 3D-mapped video projection that transformed the main façade of the city's Hallgrímskirkja church. Celebrating exceptional aspects of Arctic nature and culture, the multi-disciplinary project team was led by Marcos Zotes (UNSTABLE).

Technology | 3 x 12K lumens projectors, MadMapper software.

Magic Realism

Screenplays of Dreams
Ocubo (Nuno Maya, Carole Purnelle)

Various locations

Portuguese light artists Nuno Maya and Carole Purnelle use contemporary projection-mapping systems to express visually the Latin literary genre of magic realism: a concept whereby magic, memories, premonitions, apparitions and other aspects of imagination can merge with reality.

This page | *Sky Machine*, Toruń, Poland: In the home city of Renaissance visionary Nicolaus Copernicus, Ocubo video-mapped a multimedia projection across the Holy Spirit Church, reprising the idea of a time machine (a sky machine) spanning classical European culture and Copernicus's controversial sun-centric space science discoveries.

Opposite top | *Sintra Garrett*, Sintra, Portugal: The National Palace was transformed with 2D and 3D animations, light effects and multimedia illustrations, accompanied by original music and a reading of the Romantic Portuguese writer Almeida Garrett's poem 'Camões'.

Opposite bottom left | *The City of My Dreams*, Amsterdam, The Netherlands: Optimistic visions for Amsterdam's future, painted by young children at the ASVO school, were digitized as an 8-minute video, map-projected across the classical stone façade of the Hermitage for Children.

Opposite bottom right | *Projecting April*, Lisbon, Portugal: Dramatic archival stills and videos, mixed with newly created animations, revived the hours of Portugal's 25 April 1974 pro-democracy 'Revolution of Carnations', when enlarged to architectural scale. For the 30th anniversary celebrations in 2014, these images were 3D-mapped as a 270° projection on several adjacent buildings and structures at Terreiro do Paço, Lisbon.

Technology | These works were produced with different arrays of 20K lumens video projectors, mainly Panasonic XL 47.

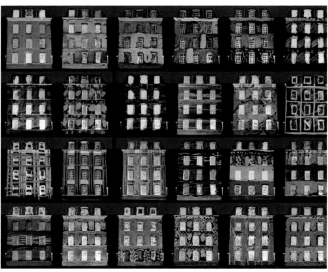

Elevations

Projecting Fantasies

086

Mixed Messages

Urban Projections
Philipp Geist

Various locations

Berlin-based electronic artist Philipp Geist creates large-scale immersive light and video art projections for urban places and art spaces. These are conceived as dynamic, provocative and anti-chronological experiences involving humans, historic and contemporary architecture, city spaces, ambient sounds, composed music, typographies and abstract imagery, all 'painted' with digital technologies like a 21st-century palimpsest (layering of diverse messages over time). Since 2005, he has performed a series of architecturally mapped video projection works in different cities, all expressing time-related concepts.

Opposite page | *Time Drifts*, Montreal, Canada.
This page top | *Time Drifts*, Frankfurt, Germany.
This page centre | *Lighting Up Times*, Barfüßerkirche Erfurt, Germany.
This page bottom | *Construction–Deconstruction*, Otterndorf, Germany.

Technology | 3-5 x 5-15K video projectors, depending on location.

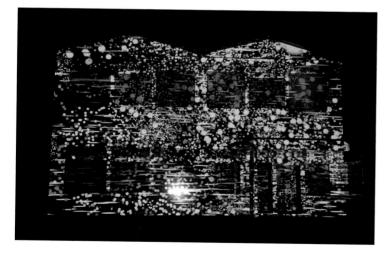

Industrial Drama

Bow West Olympics VIP Entry

Speirs + Major

London, UK

Washed with spectacular LED floodlighting, the Bow West concrete batching plant in Stratford was the iconic, industrial-era gateway for VIP guests attending Olympic Park events in London during the 2012 Games. Briefed by the Olympic Delivery Authority to follow branding guidelines specifying purple as a key colour, British lighting consultants Speirs + Major, with architects Allies and Morrison, used dichroic glass filters and wide beam metal halide floodlights, programmed to create soft, yet dramatic colour transitions. In addition, the floodlights were installed to generate shadows and highlights, emphasizing the sculpturally significant elements of the architecture. When the Olympics ended, the concrete plant was returned to its owners to resume regular operations.

Technology | 13 x 400W and 7 x 150W Enliten wide-beam metal halide floodlights and 6 x 575W narrow-beam projectors, all with magenta dichroic glass filters, 3 x 36W T8 fluorescent batten luminaires with magenta gel filters, timer control system.

Euphonic Optics

Visual Piano
Kurt Laurenz Theinert

Lüdenscheid, Germany

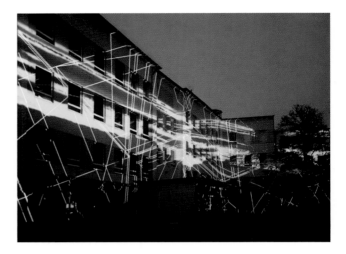

Every note of photographer and light artist Kurt Laurenz Theinert's MIDI keyboard generates digital code that causes graphics to swirl across the surfaces around his performance zone. Audiences are immersed in, and entranced by, these sound-elaborated 'drawings with light'. The video effects are generated live, from the keyboard and pedals of the 'Visual Piano' MIDI console, via bespoke software on a nearby laptop. In Stuttgart, software programmers Roland Blach (v1.0) and Philipp Rahlenbeck (v2+) worked with Theinert to develop a 'serious entertainment' system that would significantly advance post-18th-century 'colour organ' and 'light organ' demonstrations beyond the techniques that have been common among VJs in nightclubs. This 'Visual Piano' system, which Theinert and collaborators have demonstrated at venues and light festivals around the world, is unusual in its capacity to 'play' immediately visuals that directly arise from the keyboard. Beams of white light deform, scan, overlap, become colourful and throw out nets in instant, abstract responses to the music. These projections directly transform each performance environment into a hypnotic domain – a different tactic from using screens to take audiences into pre-produced realms of fiction.

Technology | Modified MIDI keyboard controller, pedals, bespoke software (version 2.0+ based on vvvv).

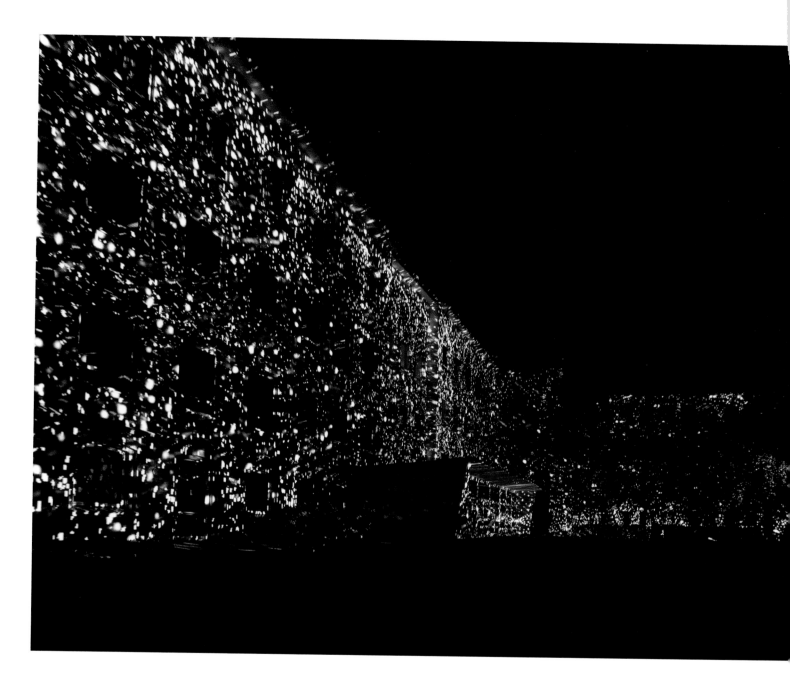

Elevations

Projecting Fantasies

Guerrilla Dada

Columbus 2.0
Michael Bielicky, Kamilla B. Richter

Seville, Spain, and other locations

Opposite page | Conceived for the Seville 2008 Biennale to highlight Christopher Columbus's voyage to America in 1492, *Columbus 2.0* was a video projection by Karlsruhe-based electronic artists Michael Bielicky and Kamilla B. Richter, symbolizing today's adventurous navigations across oceans of digital data. It involved dynamically visualizing relentless waves (an ocean) of keywords mined from Google News, using custom-programmed software.

Technology | Custom steering-wheel interface and visualization program using real-time data, portable video projectors, video splitter, laptop.

Falling Times
Michael Bielicky, Kamilla B. Richter

São Paulo, Brazil, and other locations

This page top | Luminous white pictograms flow down one wall of a São Paulo office tower. These are visual ciphers representing keyword frequency data mined from CNN broadcasts and other 'Infotainment' bulletins distributed to a global 'InfoSociety' of 'InfoConsumers'. Intended to raise concerns about social corruption caused by 'news' pollution, this video projection was first performed six months before the 2008 global financial crisis.

Technology | Several video projectors, video splitter, custom-programmed software using a real-time generated data flow.

Organic TV
William Latham

Brighton, UK

This page bottom | William Latham, a British pioneer of evolutionary biological computer art, reprised one of his classic punk-era visualizations, *Organic TV*, as a one-night, 'drive-in cinema' façade projection celebrating his 2013 'Mutator 1 + 2' exhibition at the Phoenix Gallery in Brighton. During the Brighton Digital Festival, projectionist Nick Fenwick showed Latham's historic footage from a van-mounted video projector onto a textile screen hung across the gallery entrance.

Technology | 1 x Sanyo 7500 lumens HD DLP video projector, laptop, bespoke software.

Time Blast

The Irreversible
Norimichi Hirakawa

Ljubljana, Slovenia, and other locations

Every second of time is lost with a strident clang; every
explosive vision of white star constellations in endless black
space evaporates after precisely two seconds. Japanese artist
Norimichi Hirakawa highlights the relentless, inescapable
passing of time with 1,024 colourless still images projected
on walls as a video sequence. His key message is that real
time cannot really be subverted by 'realtime' technology:
'The movie seen today does not look the same as the same
movie seen yesterday.' *The Irreversible* was first shown at the
Kapelica Gallery in Ljubljana in 2010.

Technology | 1 x DLP projector, 2 x speakers, 1 x computer.

Elevations   Projecting Fantasies

Light, Architecture and Branding

Transforming Luminous Pixels into Brand Experiences

Thomas Schielke

Abu Dhabi's Yas Island Marina Hotel includes a steel and glass canopy with over 5,000 custom LED lamps.

High- and low-res light façade of the ION shopping centre on Singapore's Orchard Road.

Global companies often exploit architectural icons to lend physical form to their desired reputations. After twilight, the natural qualities of buildings must be supplemented by architectural lighting, and today's technologies offer new potentials for luminous storytelling as nocturnal branding.

Media façades, a late 20th-century addition to the ancient typology of luminous walls, have changed the roles and concepts of architectural illumination. Constantly flickering instead of static and glowing, they offer companies vast potential to express dreams, associations, atmospheres, experiences and narratives. Without dynamic lighting and especially televisuals, architectural form serves mainly as an orientation focus.

The colour-changing net of Abu Dhabi's Yas Island Marina Hotel has vividly demonstrated how a media façade can highlight a luxury destination and help catalyze and symbolize a new economic network: in this case a Formula 1 hub in the Middle East (see also p.27). Along Orchard Road in Singapore, several shopping malls competitively express their 'spirit of place' via customized creative content for their media façades (see also p.50): this happens because their owners want to project their venues as unique experiences and have been dissatisfied with their customers' lack of awe at commercial video billboards.

Precedents of light as advertising

Nocturnal building illumination now exceeds simple visibility via floodlighting or the basic practicalities of orientation or advertising with illuminated signage. Luminous patterns on façades can act as signs to identify a specific brand and differentiate it from competitors. Smart interactive lighting concepts have become essential for commercial storytelling. Light is a complete and compelling visual language which valuably influences the minds of consumers.

Early advertising messages were communicated by illuminated billboards and frequently featured company logos, most blatantly around New York's Times Square since the 1910s. Over time people began objecting to obtrusive illuminated signs on top or in front of buildings, compromising their architectural character. This is one aspect of the 'decorated shed' phenomenon that American architects Robert Venturi and Denise Scott Brown highlighted in their 1972 book, *Learning from Las Vegas.* New York authorities now require large illuminations on all commercial media displays, necessitating owners to invest in 'bigger and better' to protect the area's global magnetism in competition with other bright light zones such as Tokyo's Shinjuku district, or the gaming strips of Las Vegas and Macau.

A more subtle approach to branding with architectural lighting involves abstract light patterns on buildings. Another is to blend lighting with the natural character of the architecture to produce a more holistic visual identity. Architectural lighting includes numerous examples where the illumination has been carefully integrated with the building.

Several music groups have used advanced stage lighting techniques to create audio-visual brand experiences. Examples include Jean Michel Jarre's 1990 La Défense concert in Paris, with spectacular laser and light projections on several tall buildings; U2's 1997 'Popmart' concert tour, with the first large flexible and foldable LED video screen; and Coldplay's 2012 'Open Air' tour, where the band controlled LED wristbands worn by many of the audience. Experimental light concepts launched at public events have often become prototypes for later illuminations of architecture.

The software-adjustable light ornament

Modern approaches to architectural lighting were dramatically introduced with the Eiffel Tower in Paris. This marvel of iron engineering innovation was launched with the first Paris Exposition in 1889, where many pavilions were illuminated by gas and some early electric lamps. Coloured lamps created pixel light patterns with basic graphic elements and introduced a layer of luminous narrative on the architecture.

After the Second World War, American lighting design leaders such as Richard Kelly, Howard Brandston and Lewis Smith focused on the qualitative aspects of lighting and aesthetic effects in interpreting space with light. Richard Kelly based his lighting strategies on three components, which he called 'focal glow', 'ambient luminescence' and 'play of brilliants'. Focal glow draws attention to an object and separates important from unimportant. With ambient luminescence, Kelly introduced diffuse general illumination to see information. His third concept, the play of brilliants, referred to light as a medium of stimulation and excitement. Stanley McCandless, an expert in stage lighting, focused more on the dramatic effects possible with supplementary lighting: to maximize visibility, emphasize forms, create complex compositions with focus, depth and perspective, or to generate mood. In stage lighting, illumination provides additional layers of meaning that help tell the story. It can create intense emotional effects in short time frames. It can even dissolve static space.

In architecture, dynamic luminous embellishments of façades, combined with digital light-control systems, have transcended the conventions and artistic limitations of physical buildings. Many media façades seem to envelope the architecture like luminous wallpaper, suggesting a new type of architecture – the building as a software-adjustable light ornament. Illuminated imagery dematerializes daytime patterns and creates a new nocturnal scenography, in styles ranging from modest highlights to virtual shatterings of the physical masonry – a popular trend in projection mapping onto historic edifices.

Venturi and Scott Brown claimed that architectural lighting could lead to non-architecture. 'These electronic elements promoting flexible imagery – graphic, narrative, abstract, and/or symbolic – work as sources of ornament that appeal to the hype sensibility of our time and as sources of information, dynamically complex and multicultural... Here architecture becomes non-architecture,' wrote Venturi.

Richard Kelly, New York pioneer of modernist architectural lighting.

Las Vegas (seen here in the 1960s) inspired *Learning from Las Vegas*, a 1972 urban design manifesto by US architects Robert Venturi and Denise Scott Brown.

William J. Mitchell, an Australian academic who (with Media Lab founder Nicholas Negroponte) led many of MIT's digital architecture innovations until the mid-2000s, noted convergences between architectural lighting design and computer graphics, because digital systems allow anything that lights up to be treated as an addressable, programmable pixel. He was sceptical of these transitions when he wrote in 2005: 'The uses of the new medium remain in an archaic, skeuomorphic phase – much like that of Greek marble temples that imitated the forms and details of their wooden predecessors, or bronze axes that replicated the leather binding patterns of wood-handled stone weapons. We are still seeing horseless carriages, wireless telegraph thinking.' Mitchell's style of critical thinking is essential for marketing and design experts to avoid the traps of imitation.

While classic architectural ornaments, such as capitals of columns, were inspired mainly by plants and other forms of nature, the ornamentation of early media façades was constrained by their reliance on electrified planes of rectangular pixels. Early screen walls were inspired by window illumination concepts or the architectural composition of each building. Later façades using LED light sources were conceived with the Cartesian geometry approach of points, lines and planes. To avoid the technical look of bare LED points, programmers more recently developed different 3D shapes for pixels, including discs, lozenges and crystals, and are debating new uses for pixels. In addition, archaic uses of basic colours are now evolving towards more sophisticated combinations of shades and colour specification techniques.

Explicit signs versus implicit symbols

The imagery of lighting for brand communication spans from explicit signs to implicit symbols. Corporate colours appear vivid but one-dimensional, while more complex light compositions enable more abstract symbolism.

When architecture inspires lighting concepts, dual contrasts of the spatial and structural compositions are often emphasized: for example, vertical vs horizontal, looking in and looking out, foreground against background, small contrasting large. But as well as the architectural perspective, lighting may originate from the desired messages to onlookers. For example, one dimension of identity, which could be addressed with light pattern, is the contrast between natural and technical qualities. These values could be projected, directly or indirectly, via light patterns and/or light installations.

When companies want to emphasize dynamic energy in their branding, they often opt for animated façades as a key communications tool. Despite the genius of many architects (including Francesco Borromini and Frank Gehry) who have created impressions of dancing structures, buildings remain inherently static. But with virtual patterns of light, architecture can be a canvas to tell explicit stories over time. Contemporary light technologies can deliver 3D impressions on any 2D façade to enhance 'brand magic'.

Lighting designers and architects have been learning from fashion stylists various ways to wrap smart luminous fabrics around buildings. This trend began with flagship stores for luxury apparel brands like Louis Vuitton and Prada, and now has reached low- and mid-budget chain stores like Esprit and H&M.

From social milieu to brand personality

When visual identities are assigned to brands and projected to target groups, marketing and sociology experts use different models to evaluate the effectiveness of the

Illuminated façade of the Louis Vuitton flagship store in Hong Kong (2005).

communication. The classic sociological model measuring from low to high class has been extended to include style as a parameter ranging from traditional to modern. Meanwhile, marketing specialists tend to personify visual identity as 'brand personality'. This concept, mostly used for product and package design, also applies to architecture and illumination. Brand personality includes properties such as temperament, competence, attractiveness and naturalness. These classifications allow differentiated evaluations to be used to analyze different lighting concepts in terms of effective brand communication.

Brand 'essence', another concept documented in corporate identity manuals, can also be a starting point for developing lighting solutions, where relevant properties include visibility, distinctiveness and consistency. A semiotic perspective could further facilitate analysis of architectural lighting concepts for effective branding.

Innovative lighting concepts for vibrant and high-tech identities

Brand-effective architectural lighting combines innovative ideas and strong design solutions with appropriate technology. An essential first step to creating a consistent and convincing brand identity is knowledge of the business. Companies often strive for an individual design to set themselves apart from competitors. But offering a clear brand story is much more important than simply attracting attention by being different. Dynamic light installations allow designers to adapt brand scenarios and stories over time and to play with tempo to achieve, for example, a spirit of vitality or temperament. Surprising light effects can create magical impressions. Examples include changing fields of colour or incorporating small logos noticed only at the scale of façade details (Louis Vuitton Hong Kong, 2005).

Telling authentic luminous stories

Lighting concepts must appear authentic to support a brand's overall identity effectively. For example, garish light sequences are appropriate to promote computer games, but are inconsistent with the respect needed by a financial institution. An example is the Dexia Tower in Brussels (2006), its constant colour dynamics contrasting with the restrained and mysterious effects on the Commerzbank headquarters (Frankfurt, 2000). A lighting solution might be distinct and impressive to gain attention, but, if it does not clearly contribute to the essence of a specific brand, it fails the basic purpose of lighting for brand communication.

Another brand design challenge relates to the character of the local neighbourhood and the factor of time. For example, Frankfurt's Zeilgalerie was one of the early shopping malls with a media façade (1992), but after about two decades the owners decided to upgrade to relate more to the central district's new 21st-century character, which has evolved with recent examples of eye-catching glass architecture. Imagery for the new Zeilgalerie (2011) replaced the former coloured media façade with a pattern language of plain white light, to balance the variety of other optical stimuli in the neighbourhood.

Recent advances in smart technologies have made luminous storytelling much easier. The key innovation of course is small LEDs, which demonstrate much higher luminous efficiency and durability than conventional light sources, as well as low operation and maintenance costs. LEDs in red, green and blue allow infinite RGB mixes of light colours for individual designs, and the small size of LED pixels has enabled new constructive solutions for delicate façade designs.

LED media façade of the Dexia (now Rogier) Tower in Brussels (2006).

Essay

Light, Architecture and Branding

Optically precise hardware systems are able to minimize light pollution to help comply with the accelerating introduction of dark sky regulations from local authorities. Also video software is more often replacing conventional light-control systems to create complex light patterns. 3D video-mapping software now allows light projectionists to match their imagery precisely with any building façade for astonishing visual effects that can mesmerize crowds at festivals and night celebrations.

Beside these control systems, interfaces for social-media content have opened up the possibility for influencing lighting effects with real-time information and user-generated content. Public observers can actively influence the content of commercial brands, depending on the level of access that the company desires or tolerates. However, the technology reaches its limit when it is mainly a demonstrative posing for size and state-of-the-art digital infrastructure, because this competition will be lost within weeks to the next project with a bigger gesture and more modern technology. For example, chemical company Bayer lost substantial money and credibility when it decided to promote its brand with an animation on the media façade of its 122m-high (400 ft) 1960s headquarters tower in Leverkusen, Germany. While Bayer's video has been quite successful on YouTube (over 100,000 views since 2009), technical problems prevented the LED installation from operating on the building.

Small interventions to mass-market initiatives

Scales of urban luminous storytelling range from the windows of semi-public pop-up stores to major urban programs to transform permanently the image of a city or region. One novelty is the interactive shop window, where motion-tracking camera systems help produce light effects to induce people on the sidewalk to interact with the systems viewing them from behind the glass.

Brands such as Diesel (Berlin, 2009), Nordstrom (Seattle, 2011), Adidas Neo (Berlin, 2012) and Nike (Selfridges London, 2013) have used these playful technologies to attract attention and start dialogues among players and spectators. These installations, where lighting has been included as part of the storytelling, express 'cutting-edge' brand messages that are highly magnetic to young people and early adopters of new technology. Diesel used severe weather situations with lighting as part of their 'Destroyed' campaign. Nordstrom allowed passers-by to write with light on the back wall of their store window. Adidas Neo connected their store window and façade with a smartphone and QR technology and allowed participants to control a virtual shopping bag or to influence the colours of the façade lighting. Nike featured reflective jackets with strobe lighting towards the street.

Temporary urban screens

Large audiences are attracted to animations on big screens and façades. While media façades are permanent installations, where the dynamic images represent an integral and original part of the building, temporary video-mapping installations can be more useful to introduce a new brand identity.

For Thorsten Bauer, head of the German design studio Urbanscreen, this difference is essential. He suggested:

The temporary installation surprises with a new interpretation of the old. ... the old is the dramaturgical starting point of the production... The façade with a permanent installation is under the pressure that it has no real identity which it can contrast with.

Ephemeral light projects are often linked to events such as anniversaries, product launches or current marketing campaigns. Examples include performances by Coca-Cola (Atlanta, 2011) and Nokia (London, 2011).

Coca-Cola impressed its followers with the world's tallest and brightest projection for its 125th anniversary at its Atlanta headquarters. The event included narrative and cinematic elements and real-time technology to integrate live social-media components on all four sides of its skyscraper. The imagery involved computer

animation, live action photography and trompe l'œil techniques supporting its multi-faceted brand identity. The storyline linked advertising from Coca-Cola's history with impressions of contemporary consumers.

For its Lumia live product launch, Nokia collaborated with the Canadian DJ and dance music producer Deadmau5, creating a spectacular show projected onto the Thames façade of the Millbank Tower in London. With over one million hits on YouTube, Nokia's marketing strategy was highly successful, particularly in building a stronger relationship for the brand with the young and fashionable target group. The show opened with the original architectural pattern of the building, changed to deconstructive transformations and equalizer figures, then finally released the product image. Nokia's Adam Johnson said: 'We wanted to do something as innovative as our products to celebrate the launch of the Lumia range. The show took its inspiration from the vibrant colours of the range.'

Permanent light scenarios

Permanent façade installations allow continuous and flexible storytelling beyond the constraints of a single event. Even if luminous imagery is dynamic, a luminous screen is a fixed element of a static building, enabling the architecture to transmit brand messages day and night.

At some buildings the screens appear similar to large televisions because of their sizes, forms and pixel resolutions. In other installations the designers have integrated media elements in more abstract ways: for example, including the screen integrally with the architectural design, or wrapping the entire building with a media façade. Cultural institutions like Austria's Kunsthaus Graz (2003) or the Ars Electronica Center in Linz (2009) exploit their media façades as laboratories to discover and display futuristic aesthetics and technical possibilities. The Kunsthaus Graz began with a white monochrome display with visible compact fluorescent lamps on one façade, while the Linz centre has colour-changing media walls on several façades, including an array of LED pixels behind glass screening.

A remarkable early colour marketing concept originated from the sport industry with the Allianz Arena (Munich, 2005), where a blue or red luminous façade announced the corporate colour of each of the two local football clubs and white highlighted the guest teams. Insurance group Uniqua has also installed media façades at two sites (Vienna, 2004; Budapest, 2009) to enhance the relevance of its brand image to trend-responsive consumers.

The strongest advances, however, especially for colour changing and media façades, come from retailers aspiring to brand images that are exclusive, luxury and vital. Examples include the Galleria Department Store (Seoul, 2004), Galleria Centercity (Cheonan, 2010) and Louis Vuitton stores across various shopping capitals.

Light festivals for city branding and development

As well as single commercial brands or institutions, communities have discovered the potential of lighting for city marketing. Today's most prestigious light festival, 'Fête des lumières' in Lyon, dates back to a 17th-century Catholic procession. Over time the light ceremony has developed into a professionally organized festival, attracting approximately one million visitors per night (according to a 2010 report by the Lyon-based Lighting Urban Community International [LUCI] network).

However, some festivals happen logically in towns where major lighting manufacturers have a long local history: for example, 'GLOW' in the Dutch city of Eindhoven, where Philips is headquartered, or 'LichtRouten' in the German city of Lüdenscheid, where ERCO is prominent.

Some economically powerful cities connect their light festivals to international conferences and/or industry trade shows: for example, the commercially managed 'Luminale' in Frankfurt, which has become the world's largest annual light event.

For various cultural and historical reasons, European cities lead the early 21st-century movement in urban light art festivals. As well as large events administered by city authorities as part of their annual tourism calendars, there is an informal network of organizers and participants in smaller city festivals, led by German light artist Bettina Pelz. In Britain, the Artichoke Trust, led by Helen Marriage, produces 'Lumiere' festivals in Durham and Derry.

ERCO's P3 building in Lüdenscheid, Germany: a monochrome treatment.

Frankfurt's Zeilgalerie shopping centre after its light façade revamp in 2011.

Across the planet, Sydney hosts 'Vivid', the world's second largest light festival, celebrated for its spectacular projections by celebrity artists on the Sydney Opera House. Founded in 2009 via state government backing of a local team led by Mary-Anne Kyriakou, 'Vivid' was catalyzed as 'Smart Light Sydney', 'the world's first eco-ethical outdoor light art exhibition', where 'carbon-belching' fireworks, flames, sky lasers, incandescent bulbs and other suspect technologies were avoided. Two further 'smart light' events were curated by Kyriakou's team at the Marina Bay waterfront renewal zone in Singapore in 2010 and 2012.

Recovering dark skies

Contradicting branding initiatives where light, colour, dynamic light scenes and high-tech interactive installations are considered positive contributions to urban development, some communities and regions have declared darkness to be an essential aspiration to support environmental solutions. As well as the globally successful 'Earth Hour', a 60-minute community power switch-off promotion run by the World Wildlife Fund for Nature, dark sky reserves have been declared in Canada (Torrance Barrens Conservation Reserve, Ontario, 1999), Britain (Northumberland, 2013) and Germany (Gülpe, 2013). Environmental organizations and astronomers generally seek reduction of light pollution (photon wastage) to achieve a dark sky and clear views of the stars: they are influencing the down-directed and up-shielded designs of new LED streetlights.

Smart techniques for urban lighting

Both architectural lighting and media façade systems have specific design parameters that can be manipulated to enhance brand communication. An obvious example is light colour, which can be linked to corporate colour schemes: for example, petrol stations with illuminated signs along their roof edges.

Adjusting lamp brightness allows companies to increase visibility of their branding. Perhaps more important is the overall quality of an installation, where the luminous pattern reveals an indirect connection to the brand personality. Too-bright lighting could reduce a brand's credibility on influential credentials such as environmental awareness and naturalness.

When creating media façades, the number, size and form of pixels are essential design criteria. The expression of a pixel could range from unadorned technical light 'dots' to various artistic shapes. High-resolution screens can show TV-like content with highly detailed images. However, that approach can be aligned too closely with conventional TV commercials, which are considered to lack the artistry required for branding campaigns. Low-resolution screens are not used for quotidian advertising, so can be effectively combined with high-res screens (Wilkie Edge, Singapore, 2009).

Important concerns when integrating video screens with architecture are the daytime appearance of a building and the effects of direct sunlight. Most media façades have been designed to create an impressive image at night and benefit from the low luminance of the dark sky. But the daytime look and legibility are also crucial for a successful design, especially with large screens. For this reason, screen manufacturers have introduced specific techniques to offset the effects of direct sunlight and overcome the challenges of luminance contrast, to enable good visibility.

From scene to sequence

Lighting parameters such as colour, brightness and pattern could be included in a scenography that changes over time – either as a pre-programmed sequence or in response to sensors or other interfaces. However, simple changes of colours throughout a complete colour space do not tell a specific brand story. Such a strategy might even be considered generic if nearby buildings tell a similar story with rainbow sequences. It could also reduce observers' orientation when a building is seen from afar in one colour but changes to another identity on approach. To achieve clearer identity and recognition, some architectural light installations have been schemed in just one colour (ERCO P3, Lüdenscheid, 2002) or just white (Zeilgalerie, Frankfurt, 2011).

Networking light with sensors and smartphones

Impulses controlling light patterns can be triggered by pre-programmed light scenes or signals from sensors or mobile interfaces. Motion, sound or temperature sensors have enabled direct or indirect relationships between buildings and their environments (Zeilgalerie, Frankfurt, 1992). Public participation is possible when a light system allows connections to mobile devices so operators can actively communicate with the brand. Mobile phones can send text or image messages to link a light system to social-media platforms or to instruct specific functions. Smartphones with a camera can also activate luminous façades via a QR code-scanning app linking to augmented reality content (N Building, Tokyo, 2009).

To help compensate for the energy demands of increasing numbers of media façades, photovoltaic systems can convert sunlight into nocturnal illumination (GreenPix, Beijing, 2008). Especially for temporary video-mapping installations, special software can be used to create stunning transformation effects that play with the real architecture and virtual building patterns.

Luminous storytelling

As related genres, interactive light installations and media façades have evolved from temporary artistic installations into permanent interfaces for commercial branding and to encourage consumers to shop. Here the value of architectural lighting is defined by its semiotic function as well as its scenographic and storytelling qualities. Where customers are annoyed by traditional billboards or are in locations where conventional advertising is restricted, architectural lighting can allow companies other promotional potentials. With a heightened and expanding community awareness of environmental issues, architectural lighting and media façades are obvious targets of public debates on energy and sustainability ethics.

With their powers to facilitate impressive imagery, media façades have dissolved many earlier differences between architectural lighting and motion graphics. Influences from MTV videos – such as cuts, transitions and switches of motion speed – have been obvious in many recent façade animations. Whereas some projects have appeared as 'decorated sheds' with dynamic pixelated ornaments, other buildings have displayed the more creative approach of merging architectural hardware and light-control software.

Authentic branding requires that buildings with striking façades also offer interiors of congruent quality – and that tenant companies purvey products and/or services of similar standard. Social-media and video platforms are now essential channels for sharing luminous brand experiences with not-there audiences. They link place-fixed installations with the virtual world to enable – and demand – comprehensive brand communication.

Luminous Structures

A Timeline of Historical Triumphs

Thomas Schielke

From the stained-glass tableaux of medieval French cathedrals to giant screens of solar-powered pixels flashing across today's Asian cities, luminous structures are history's nexus between the arts of light and building.

This chronology shows notable creative advances during the three great eras of lighting technology: natural radiance, electricity, and semiconductor-enabled luminescence.

1370-1700 **The Registan, Samarkand**
Three medieval civic buildings in this Silk Road city dazzle with decorative, turquoise-glazed terracotta tiles and gilded mosaics.

1291 **Nanzen-ji, Kyoto**
Translucent shoji paper walls reduce daylight brightness and transmit a lantern effect when viewed externally at night.

1653 **Taj Mahal, Agra**
A radiant complex of white marble structures around a domed mausoleum, inlaid with jewels, and reflected in nearby pools.

1248 **La Sainte-Chapelle, Paris**
Extensive use of stained-glass panels create decorative lantern effects when the cathedral is internally illuminated at night.

1648 **Bayt al-Suhaymi, Cairo**
The main façade of *mashrabiya* (carved wooden lattice) screens creates a lantern effect when viewed externally at night.

1802 **Soho Foundry, Birmingham/ Smethwick, UK**
First exterior building illumination using coal gas lamps, demonstrated by William Murdock.

Natural Radiance
Sunlight filtering or reflections, non-electric lamps

1931 **Kansas City Power & Light Building, Kansas City**
Multi-colour flickering floodlights on terraces, red-orange 'fire' lantern at top.

1910 **Gas and Electric Building, Denver**
The façades were fitted with an array of 13,000 incandescent bulbs.

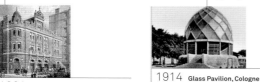

1908 **Singer Building, New York**
Roof outlined with electric lights; tower floodlit from base.

1931 **Empire State Building, New York**
Mast lit with reflector lamps; 1945 mast interior lit; 1956 beacon lights added; 1964 floodlights/colour gels; 2012 LED system.

1889 **Eiffel Tower (Paris Exposition), Paris**
Hundreds of gas street lamps were arrayed around the tower, while lamps on a circular rail projected red, white and blue beams from the beacon.

1921 **Wrigley Building, Chicago**
Entire tower floodlit from its base, topped by a revolving beacon.

1914 **Glass Pavilion, Cologne**
Faceted glass exhibition pavilion with 14-sided base of glass bricks: a giant jewel/prism reflecting coloured rays from the sun.

1881 **Savoy Theatre, London**
First public building lit by Joseph Swan's incandescent carbon filament lamps: 1,200 bulbs powered by a generator.

1929 **German Pavilion, Barcelona**
Mirrored columns under a 'floating' roof, with fluorescent backlighting inside a milky glass light box, pools, glass and polished stone walls.

1928 **De Volharding Building, The Hague**
Corner building of glass sheets and blocks, with a light tower and advertising signs: early example of 'luminous architecture'.

1878 **Grands Magasins du Louvre, Paris**
First use of 'Jablotchkoff candles', 80 carbon arc lamps powered by AC electric current: part of a city-wide installation program.

1913 **Woolworth Building, New York**
Mazda C nitrogen floodlamps in mirror reflectors were set to increase light intensity from levels 31 to 60, topped by a 'ball of fire'.

1927 **Edison Building, Philadelphia**
Multi-coloured floodlights individually dimming and fading, and a rotating beacon.

1950 **860-880 Lake Shore Drive, Chicago**
Twin apartment towers with glass curtain walls and backlit glass walls in the lobby.

Electricity

Heat-generating lamps, frequent use of glass bulbs and windows, mechanized motions

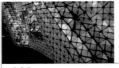

2000 **New 42nd Street Studios, New York**
Programmable theatre fixtures project coloured lights on a glass façade with a Light Pipe tower and perforated metal screens.

1999 **Aegis Hyposurface, Birmingham**
Algorithmically controlled pistons pump waves of undulations across a faceted metal screen, with lighting projection.

1999 **Marnix 2000, Brussels**
Bank building façade as video screen, playing animations uploaded by citizens to a website.

1963 **Beinecke Rare Book and Manuscript Library, New Haven**
Glowing amber 'jewel box' with backlit marble façade walls set into grids of sculpted steel mullions.

1990 **La Défense, Paris**
Projections on monuments, electronic music and fireworks to celebrate the 200th anniversary of the French Revolution.

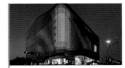

2004 **Galleria Centercity, Seoul**
Department store façade with 5,000 programmable LED discs, creating motion graphics at night.

1960 **Thyssen-Hochhaus, Düsseldorf**
Steel and glass office building in 'three slices', 95m (310 ft) high; two alternative fluorescent light scenes show horizontal lines or logo.

1987 **Institut du Monde Arabe, Paris**
Glass façade over an ornamental screen (brise-soleil) of 240 photo-sensitive metal shutters designed to auto-filter daylight.

2000 **Expomedia Light-Cube, Saarbrücken**
Cubic building with metal mesh solar skin; coloured LED lights projected on the screens at night.

2003 **Kunsthaus, Graz**
BIX (big pixels) array of compact fluorescent lamps on curved skin of building; can be programmed by artists.

1958 **Seagram Building, New York**
'Tower of light' from warm white ceiling fluorescents, lobby lit with PAR lamps grazing down gleaming travertine walls.

1986 **Tower of Winds, Yokohama**
Exhaust outlet with perforated metal skin, backlit by a digital system of coloured lights displaying environmental data (e.g. noise).

2000 **Olympic Boulevard Light Towers, Sydney**
Nineteen 30m-high (100 ft) lighting pylons with 4 sq m (43 ft²) glass mirrors and translucent glass solar collection.

Electroluminescence

LEDs, sensors and other semiconductor systems, cool-touch lamps and screens, computerized dynamics

2006　**Dexia (now Rogier) Tower, Brussels**
Office tower with 12 RGB LED lamps at each of 4,200 windows; programmable façades, sometimes with public participation.

2006　**Uniqa Tower, Vienna**
Continuous media screen with 182,000 programmable LED pixels inside a double-glazed façade.

2005　**Diffraction/Torre Agbar, Barcelona**
4,500 LED devices create moiré (diffraction) effects around the aluminium and glass cladding of the bullet-shaped tower.

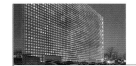

2008　**GreenPix façade (Xicui Centre), Beijing**
First zero energy LED media wall, 2,000 sq m (21,500 ft^2); photovoltaic arrays capture solar energy to power the wall at night.

2007　**Bosphorus Bridge, Istanbul**
LED floods and spots illuminate the towers, zig-zag cables and deck rails of the first suspension bridge between Europe and Asia.

2009　**77 Million Paintings/Sydney Opera House, Sydney**
Bio-fuelled projections mapped on both façades of Sydney Opera House; 'ambient' highlight of world's first 'Smart Light' festival.

2009　**Wilkie Edge, Singapore**
Advertising amplifier (A.Amp) system of low-res video patterns on Venetian blinds, relating to commercial high-res screen.

2009　**American Eagle (Times Square), New York**
25-storey façade of 12 panels, 3.3 million LED pixels, linked to store audio-visual system and images from social media.

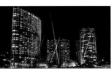

2009　**City of Dreams, Macau**
Luminous exchanges between buildings to establish an impressive landmark for an urban entertainment complex.

2013　**Sea Mirror (One Central Park), Sydney**
Cantilevered heliostat with 320 mirrors bounces daylight down inside a shopping centre; LED arrays for night performances.

2013　**The Bay Lights, San Francisco**
An animated relight of the 1936 Bay Bridge using 25,000 LED lamps with fibre optic cables clipped to 300 vertical sections.

2012　**Twilight Epiphany, Houston**
Radiant sky-viewing pavilion at Rice University lit with LEDs and designed for poetic experiences as the sun sets and moon rises.

Timeline　　　　　　　　　　**Luminous Structures**

Environments

Dead Zones, Dark Waters

Don't expect any significant city to be devoid of desolation and danger. Indeed some European capitals, especially London, revel in their histories of assaults and murder under the shroud of darkness.

Newer cities seem more sanitary – as futurist William Gibson once complained about Singapore. Barcelona architect Oriol Bohigas and Paris designer Philippe Starck dismissed Sydney's 'clean' demeanour on their 1990s visits, Bohigas remarking: 'Any really great city makes you feel there's been blood on the streets.'

Sydney recently jailed two retired detectives who murdered a student drug dealer in a suburb covered by street cameras. But these were pre-Google baddies. In most prosperous cities now, constant surveillance of public areas is deterring violence: partly because Hollywood films, YouTube videos and smartphone cameras routinely show the algorithmic feature-recognition and GPS location-tracking accuracy of sensor-fitted devices. Like prisoners in a Panopticon, we never know when our guards are remotely watching.

What does danger mean to designers of urban lighting? At one level, they must help dispel perils for people using shadowy areas. Yet outré artists often subversively want to emphasize the dark aspects of human circumstances...

Many light artists enjoy confronting witnesses with site-specific night works that exploit the weedy decay of alleys, vacant sites, derelict buildings, motorway and bridge undercrofts, car parks, road tunnels or railway cuttings.

Often these are temporary installations for festivals, commemorations or promotions. Or they could be guerrilla invasions by squads of students testing provocative ideas and novel techniques. Night-bombing town walls with portable projectors is part of the tertiary curriculum for students of Czech artist Michael Bielicky, leading the University of Art and Design at ZKM in Karlsruhe, Germany.

Many transport and infrastructure agencies see advantages in funding light artworks for potentially troublesome or visually prominent places beside busy roads. One example is artist Erwin Redl's *Passing Through Light* installation of

LED fixtures beaming bold colours up the bleak concrete walls of a motorway underpass in Charlotte, North Carolina (see p.112).

Alluring especially to light artists are the shimmering reflection potentials of black urban waterways. Witness Aether & Hemera's *Voyage* of boats floating LED 'candles' across a calm Thames inlet at London's Canary Wharf (see p.114), or the sinuous streams of LED-lit, water-filled plastic bags that Madrid's Luz Interruptus laid along cobbled lanes and pebbly shorelines licked by the ebbing tide (see p.159). In 2009, a team led by Jonathan Laventhol and Natalie Jeremijenko floated sensor- and wiki-networked LEDs on buoys in New York's Bronx and East Rivers – a concept they titled *Amphibious Architecture*.

One of the world's most successful recent examples of landscaping and lighting a derelict urban zone is the rejuvenation of a defunct elevated freight railway above New York's Meatpacking district – now a linear park called the High Line (see p.142). Lighting designer Hervé Descottes (L'Observatoire International) specified LED strips fitted to the underside of waist-high handrails along the edge of the walkway. This keeps light sources and intensity low to create a romantic mood for wanderers, emphasize the lights of surrounding buildings, and avoid contributing more photon pollution to Manhattan's night sky.

Descottes has written (with Cecilia Ramos) an influential manual of urban lighting concepts, technical charts and case studies. Called *Architectural Lighting: Designing with Light and Space*, this 2011 volume promotes precise 'sculpting' of light. It proposes direction, diffusion and density principles that are easily controlled inside photography studios but which require more skill for unpredictable outdoor conditions at the scale of the city. Descottes draws inspiration from theatre set design, noting:

The careful control of distribution and directionality of light is a lighting designer's most powerful tool in defining and revealing the limits of space. ... margins and borders of architectural bodies seemingly appear and disappear, divide and unify, guiding a continuum or break of motion through time and space.

Derelict sites and districts of cities always seem poised for reinvention. When local demographic and economic conditions reach critical points of tension, nondescript lighting must be upgraded to express the desired new spirit of place and support new surges of activity and occupancy.

Tight budgets often demand installation of versatile fixtures. In the late 1990s, Sydney lighting engineer Barry Webb and architect Alec Tzannes produced then-innovative, multi-functional 'smartpoles' to tidy the city's main streets in time for its 2000 Olympics tourist influx. Smartpoles are now ubiquitous symbols for 'smart cities' around the world.

One recent variation on smartpoles has been effective in the low-income district of St Catherine at the French port city of Le Havre. The council installed a group of 10m-high (33 ft) light poles, slightly inclining between clusters of trees around the town square. Each was fitted (above head height) with two vertical fluorescent tubes and a trio of small projectors, which can be adapted with gobo ('go before optics') colour and image filters. The poles provide not only general lighting but also perimeter definition for the plaza, navigation signals for citizens at a distance, and creative effects at community gatherings. The whole point of lighting dead areas of a city is to rescue them back for human life.

Road Trip

Passing Through Light
Erwin Redl

Charlotte, USA

New York- and Ohio-based light artist Erwin Redl installed sixty RGB colour-changing LED spot and floodlights to enliven the West Trade Street underpass of the 1-77 Bridge that divides two districts in Charlotte, North Carolina. Oval pools of bright-coloured light dramatize the coloured wall-washing of two slanted concrete embankments, and narrow spotlights illuminate the steel girders of the underpass. These lights cause intriguing shadow reflections from the concrete columns onto the bridge roadway beams. This permanent artwork, curated by Jean Greer for the Arts and Science Council of the City of Charlotte and Mecklenburg County, North Carolina, dramatically activates this once-dismal urban barrier between Charlotte's Third Ward and the Biddleville/Wesley Heights neighbourhood.

Technology | 20 x Altman Outdoor Spectra RGB LED floodlights, 20 x Chauvet COLORado RGB LED long-range narrow spots, 20 x ETC Selador D40XT RGB LED round spots, ETC Mosaic Show control system.

Environments

Dead Zones | Dark Waters

Flotilla of Light

Voyage
Aether & Hemera

London, UK, and Scottsdale, USA

Three hundred illuminated 'paper' boats float across a city canal after twilight. Igniting fond memories of childhood pleasures, this flotilla is an organic matrix of tiny light sculptures ... or it could be interpreted as a buoyant 'pixel screen'. Each 600mm-long (23½ in) boat is folded from a sheet of non-toxic and recyclable polypropylene, which diffuses light from an onboard LED. All the LEDs can be remotely operated by riverside observers, using a Wi-Fi link to the *Voyage* website. The control system involves bespoke software on a Raspberry Pi computer, controlling DMX circuitries on an Arduino board. Heavy-duty threads link the boats as a net that is anchored at perimeter points. Popular at community events in Canary Wharf in 2012, then Scottsdale, Arizona, in 2013, and other locations, the *Voyage* concept was developed by British studio Aether & Hemera (named after the ancient Greek god and goddess of bright atmospheres and daylight). Studio principals are Gloria Ronchi, a light artist, and Claudio Benghi, a media architect.

Technology | 300 x RGB LEDs, Raspberry Pi, Arduino, custom software.

Environments

Dead Zones | Dark Waters

116

Radiant Strands

Fancy/Lightweight
Cornelia Erdmann

Singapore

Delicate strands of coloured light are laced like giant fans through the pillars of waterside canopies outside Singapore's ArtScience Museum at Marina Bay. German multimedia artist Cornelia Erdmann used different colours of energy-minimal electroluminescent wire to reconfigure spatially the normally banal undercroft as a lacy maze reminiscent of a spider's web, offering wanderers a novel and fascinating interaction with the shaded space. Her installation highlighted the traditional symbolism and gestures associated with bamboo fans in many Asian cultures. They are used not only for cooling air but also as gifts and props to express respect and goodwill. This site-specific work was designed to support the eco-ethical principles of the Singapore government's 2012 'iLight Marina Bay' celebration, and was revised for the 'LUX' light festival in Wellington, New Zealand, in 2013. Carried by the artist in a suitcase, the wires and fittings are mostly recyclable.

Technology | Electroluminescent coloured wire, power inverters, cable ties, hooks.

Emergency Follies

Bibigloo
BIBI

Singapore, and other locations

This page | French artist BIBI (Fabrice Cahoreau) creates fantasy follies and scenes at light festivals around the world. The Bibigloo is his 'post-modern' (ironic) appropriation of a traditional Alaskan Inuit dome shelter. Assembled from 250 red polyethylene containers instead of blocks of ice, it measures 2.7m (nearly 9 ft) high and 4m (13 ft) diameter, and shines like a lantern from one string of fairy lights placed inside. Visitors can inspect its interior through small holes in the walls, or children might crawl through the tunnel entry. After each festival, the Bibigloo is dismantled and moved to its next location.

Technology | 1 x string of 20 fairy lights.

Le Roi des Dragons (Dragon King)
BIBI

Lyon, France, and Dubai, UAE

Opposite page | Built in 2012 for the 'Fête des lumières' in Lyon, this fantastic floating light pavilion was inspired by that year's Chinese astrology dedication to water dragons, the Lyon-favoured medieval legend of St George slaying a fire-breathing dragon, and the Egyptian crocodile god Sobek. French artist BIBI created the royal creature with 500 red polyethylene containers and 90 traffic cones. It is brightly illuminated with various kinds of LED lamps, which change colour 'according to the dragon's mood'. A large LED screen shows dramatic imagery of the four elements of earth, fire, air and water: BIBI suggests plastic as the fifth element of today's industrial world.

Technology | 1 x LED screen, Ayrton moving head LED projectors, various LED bulbs, PAR floodlights, LED tubes, DMX interface.

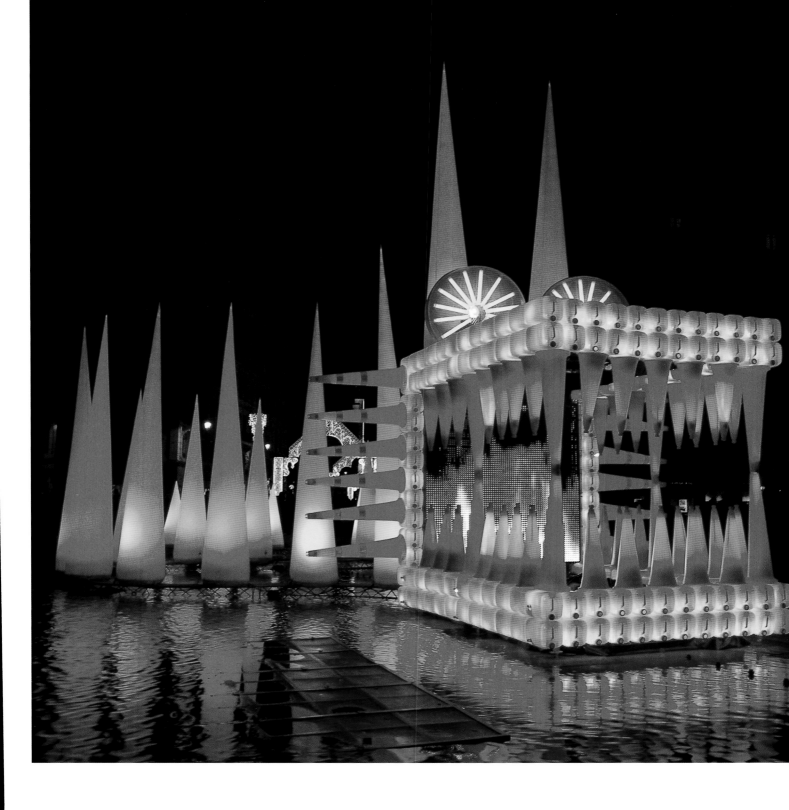

Environments   Dead Zones | Dark Waters

120

Ethereal Impressions

The Waiting
Titia Ex

Vlieland, The Netherlands

Opposite top | Inspired by a Tom Petty lyric, Dutch light artist Titia Ex orchestrated a disconcerting art installation of forty computer-animated, battery-powered objects, each fitted wtih sixteen red LEDs, all randomly blinking (long-short-long) in a dark, moon-reflective pond surrounded by bush. This was a site-specific temporary contribution to a music and arts festival on the island of Vlieland in 2010. The luminaires were fixed on sticks, radiating light from below the water surface.

Technology | 40 x computer-animated objects, each with 16 x red LEDs.

Fairy Lake Kayak
Stephen Orlando

Huntsville, Canada

This page top | A three-way network of digital devices allowed Canadian photographer Stephen Orlando (Motion Exposure) to produce magical images of coloured light dancing across the placid waters of Fairy Lake in Ontario. RGB LED strip lights were fixed along a kayak paddle and programmed to change colour via an Arduino microcontroller, allowing Orlando to capture precisely multi-coloured light trails left by the kayaker's bursts of paddling through long exposures.

Technology | Nikon D7000 camera with Nikkor 18-70mm lens, RGB LED light strips, Arduino microcontroller.

Chimney Corner #2
Vicki DaSilva

Cape Breton Island, Canada

This page bottom | In this time-exposure (film) image from New York artist Vicki DaSilva, veils of silky white gauze appear to float above a beach. In the dark, DaSilva swirled a fluorescent tube lamp to create delicate swathes of blue light. Her 2010 *Chimney Corner #2* photograph was captured on Fuji Velvia 50 transparency film, using a Mamiya 645 square format camera, at Chimney Corner Beach on Cape Breton Island, off Nova Scotia.

Technology | Fluorescent tube lamp, Mamiya 645 camera, Fuji Velvia 50 transparency film.

Tunnel Visions

Urban Green
Ljusarkitektur

Stockholm, Sweden

This page top | In Stockholm's downtown zone, a gloomy area under a bridge crossing over Kungsgatan was transformed with a temporary installation of light, sound, scent and furnishings. Ljusarkitektur (now ÅF Lighting) used the concept of 'eco ducts': frontier zones that link edges of biologically different environments, encouraging animals and plants to cross. In order to foster a new impression of security, the designers used park benches and alluded to familiar motifs: for example, the green-illuminated ceiling, emulating a meandering, mossy pathway, featured 'flowers' of upside-down green LED desk lamps.

Technology | Oco 33cm (13 in) and 55cm (21½ in) desk lamps, Philips PROflood gobo projectors.

rocklights
Ingo Bracke

Sydney, Australia

This page bottom | For the first 'Smart Light Sydney'/'Vivid' festival in 2009, German architect-artist Ingo Bracke projected colour-changing light patterns across the barrel-vault ceiling of the Argyle Cut road tunnel in Sydney's historic The Rocks precinct. His animated patterns of lines were inspired by architectural drawings, the Aboriginal idea of Dreamtime songlines and Polynesian stick charts for nautical navigation.

Technology | 2 x MSR 575W and 8 x 500W halogen projectors with different optical systems, 360° light art projection mapping software, sound system.

Human Effect
Yandell Walton

Melbourne and Sydney, Australia

Opposite page | Lasting only moments, Yandell Walton's vegetation-rich video triggers memories of pre-colonial landscapes, contrasting with the man-made stone walls on which the images are projected. The video's short duration also symbolizes biological brevities of time between birth and death.

Technology | 1-2 x HD projectors, 1-2 x Microsoft Kinect input sensing devices, bespoke software.

Environments

Dead Zones | Dark Waters

126

Motorway Signals

The Nyborg Bridges
ÅF Lighting

Funen, Denmark

Opposite page | On Funen island, Denmark, a pair of road and rail bridges across the E20 motorway were black spots for vehicle accidents, requiring improved illumination of the underpass zone. Trans-Scandinavian designers ÅF Lighting developed two strategies, combining safety, durability and creative criteria. One is a chiaroscuro (light and shadow) interplay of sepia foliage patterns across the concrete beam edges, created with 24 Philips PROflood metal halide gobo projectors. The other is an atmospheric treatment using Philips LEDline2 fittings to illuminate all the bridge pillars with glare-free green downlighting.

Technology | For each bridge pillar: 48 x Philips LEDline2 BSC713 20W LEDs and 34 x BSC716 39W LEDs; total 24 x Philips PROflood DCP608 gobo projectors (mounted on 12 poles), each with 150W CDM-T Short-Arc 4200K.

Aspire
Warren Langley

Sydney, Australia

This page | Glass and light sculptor Warren Langley's succulent 'trees' symbolize new hopes growing and glowing beneath a looming motorway which mutilated Sydney's historic Ultimo neighbourhood in the 1970s. Strengthened by steel frames and radiant from internal LED lamps, these fourteen yellow polyethylene forms appear to support the motorway like caryatids under the pediment of an ancient Mediterranean temple.

Technology | 180m (590 ft) Osram LINEARlight LED lamps, steel frames, moulded polyethylene cladding.

Projected Reflections

Q150
Ian de Gruchy, Laith MacGregor,
Kate Shaw, Guan Wei

Brisbane, Australia

When Australia's Queensland state and Brisbane city celebrated their 150-year anniversaries in 2009, the city council built a twin-tower slide projection system to embellish the multi-arched William Jolly Bridge regularly with light art imagery at night. Ian de Gruchy, a Melbourne-based international projectionist who attended school in Brisbane, was commissioned to design the system. De Gruchy's inaugural lighting display comprised twelve slide images, which highlighted Brisbane's floral emblem (the poinsettia), the city flag, imagery by indigenous artist Lilla Watson, and enlarged signatures from early city leaders, including William Jolly, the first Lord Mayor of Greater Brisbane.

This page top | *Dreamin' About a Place I'll Never See*, Laith MacGregor.
This page bottom | *Summer Solstice*, Kate Shaw.
Opposite top | *Boatmen*, Guan Wei.
Opposite bottom left | *Wattle*, Ian de Gruchy.
Opposite bottom right | *Poinsettia*, Ian de Gruchy.

Technology | 4 x Pani BP 4 CT projectors, slides mapped in Photoshop.

Environments

Dead Zones | Dark Waters

Starlight Arrays

Glühwürmchen (Glow Worms) Project
Francesco Mariotti

Various locations

Opposite page | Swiss-Peruvian light artist Francesco Mariotti is a leader of the Zurich-based *Glühwürmchen* project, for which artists and environmentalists promote landscape biodiversity by celebrating glow worms (fireflies) as nocturnal and bioluminescent indicators of healthy habitats for many kinds of creatures. Mariotti's most influential contribution to the biodiversity cause is a series of firefly-inspired art installations, which he assembles at light festivals around the world. Also to draw attention to the problem of global waste of non-degradable plastic packaging, he uses crushed PET bottles as deceptively naturalistic 'pods' for his solar-powered LED units, which glow either green or blue.

Opposite top left | *The Fireflies Factory*, Lindabrunn, Austria
Opposite bottom left | *Quantum Flowers*, Ludwigsburg, Germany
Opposite right | *The Fireflies Fence*, Marina Bay, Singapore.

Technology | 12V solar-powered LEDs (green or blue), PET bottles, cable ties, (optional) wire fencing panels.

Crystallized
Andrew Daly, Katharine Fife

Sydney, Australia, and Singapore

This page | Suspended below the bleak concrete overpass of Sydney's Cahill Expressway, *Crystallized* was a dazzling stretch of 'inverted terrain' glowing in technicolour. Created by Andrew Daly and Katharine Fife for the 2011 'Vivid' festival, the work comprised eight rectangular acrylic panels intensively perforated by 4,600 acrylic tubes, forming irregularly undulating pixel screens. Looped above this 'suspended ceiling' were 100m (330 ft) of LED striplights, powered to make the coloured rods glow with rainbow effects. Daly and Fife suggest the work can be compared either to 'a cave of stalactites or a starry night sky'. The work was also presented at Singapore's 'iLight Marina Bay' festival in 2012.

Technology | 8 x suspended acrylic panels supporting closely packed KF Plastics acrylic tubes of variegated colours and lengths, 100m (330 ft) Superlight LED strip lights.

Parks, Plazas, Promenades

Where can people wander in a city after dark? During daylight, open spaces encourage relaxation; at night they can feel uncanny, unsafe, repellent. Strategic lighting seems essential to evaporate the gloom.

City parks, plazas, promenades and lanes gain unique nocturnal qualities from their terrains, weather, air quality and other geographic conditions, and from post-twilight social activities. Any area's magnetism can be mathematically analyzed by 'Space Syntax' mapping experts, who test its integration within the local network of traffic routes and flows.

Governments of competitive cities are introducing and expanding smart lighting of zones that have potential to boost local night economies. They employ consultants to produce masterplans for public lighting of specific precincts – and ambitious authorities sponsor regular festivals of outdoor light art.

Goals of most urban lighting strategies are to illuminate and promote a city's key natural and architectural assets and zones of action. Reports also suggest ways to foster public safety, social harmony and shared local pride. They allude to local historical, geographical, cultural, visual and traffic conditions.

But intelligent lighting masterplans also emphasize the desirability of darkness, especially to support creatures (including insects) living in parks and wildlife habitats. Animals are most disturbed by non-yellow light (from white LED and fluorescent lamps, for example), uplights and luminaires with high glare or spill (where light is cast more broadly than is needed or desirable).

Visibility is the vital strategy to reduce vulnerability. However, many governments prefer to minimize spending on lighting for unused recreation areas such as tennis and basketball courts, skate parks, playgrounds and roadside picnic areas. They do not want to provide just enough light to misguide people into visiting domains that are not safe. In areas attracting crime, authorities tend to install remotely monitored cameras (including dummy cameras). Instead of programming luminaires to switch on or off at specific times, lights can be activated more efficiently by sensors detecting nearby motion or photoelectric cells that respond to surrounding levels of light.

Since the 1970s, when urban planners began to react against inhumane strategies in urban design and architecture, many local authorities have improved their policies for 'crime prevention through environmental design'. They encourage architects, planners and place managers to use passive surveillance principles, inviting people to use and watch public areas as much as possible.

Place legibility is the most basic function for outdoor lighting at night, but techniques for light clarity often conflict with strategies to create romantic moods or festive spectacles. White light, now widely available via LED luminaires for streetlamps, provides the strongest visibility – yet it can cause excessive glare and is being blamed for health disruptions among overly exposed humans and fauna. Many people retain a primitive appreciation for the apparent warmth of lights coloured yellow to red – and a converse horror of dark, long shadows across public ground. These are merely some common issues which must be balanced by designers.

Also critical in lighting civic areas are the ethics of spending public funds wisely on durable and vandal-resistant fixtures – and increasingly using solar and other renewable sources of power. Careful choices of equipment are essential for energy efficiency: long-life, high-intensity lamps, and quality operating devices such as electronic ballasts, transformers and igniters. Designers are guided by international DIN standards, national health and safety laws, and local urban planning regulations.

Beyond direct manipulation by designers is moonlight. Although the surface of the moon itself seems almost colourless, our satellite planet constantly reflects sunlight to the opposite longitudes of the Earth. Depending on the phase of its monthly and nightly cycles, and local natural conditions, its atmospheric effects on urban spaces can be magical. The moon's colouring is always strongest and warmest near the horizon, and lighter and greyer at the height of its overnight trajectory.

Planning light for public areas can be a highly sophisticated, creative and technical endeavour. In *Light Perspectives*, a 2009 survey of concepts from editors at German equipment manufacturer ERCO, 21 design choices are extensively illustrated and explained. Infinite nuances are possible within these few examples of contrasting approaches: brilliance/glare, foreground/background, architecture/theatre, looking in/looking out, diffuse/directed, neutral/expressive. As the authors (Aksel Karcher et al) noted:

Light – which includes light and dark and everything in between – is a universe of possibilities.

American modernist lighting designer Richard Kelly developed the classic concept for lighting spaces, with his 'focal glow', 'ambient luminescence' and 'play of brilliants'. Focal glow relates to an artificial version of a daytime shaft of sunlight – a strategy to highlight one or a few special features. Ambient luminescence is the overall general lighting of an area (including both horizontal ground and vertical walls), giving a diffuse and uniform level of optically comfortable illumination. Play of brilliants (not always used in simple lighting schemes) refers to a fleeting, dynamic spectacle, in which light becomes a narrative rather than constant element.

Juxtaposing ambient illumination and special effects has always been the basic (optical) test of creativity in lighting design. When travelling the world, it can be diverting to inspect night-lighting arrangements in different city spaces. Compare Barcelona's La Rambla with London's Lincoln's Inn Fields or Tokyo's Imperial Palace Gardens... Civic lighting is a realm of creativity that will reward any wanderer with a discerning eye.

Technicolour Temple

Temporary Temple Pavilion
Abin Design Studio

Hooghly, India

How to create a splendid temporary temple, on a tiny budget but with many helping hands, to celebrate the idol of an Indian village religious carnival? Kolkata-Delhi architects Abin Design Studio dazzled a 2012 festival crowd at Hooghly, West Bengal, with a technicolour matrix of tightly spaced bamboo poles, all tipped with retroreflective vinyl tape to glow mysteriously under halogen floodlighting at night. The 1,800 bamboo poles were cut in fifteen gradually varied colours and lengths ranging from 60 to 460cm (24–180 in), then planted approx 2.5cm (1 in) apart in a circular array on flat ground, surrounding a cylindrical white container for the spirit of the festival deity. After the celebration, the poles were reused as fenceposts for the village football field.

Technology | Retroreflective vinyl tape, halogen floodlighting on stands.

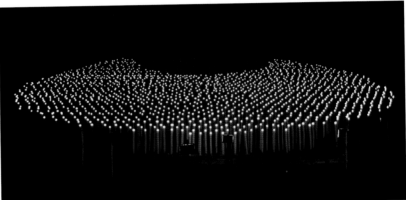

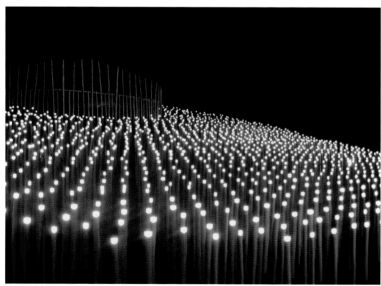

Environments

Parks | Plazas | Promenades

Sensor Responses

Pulse Park
Rafael Lozano-Hemmer

New York, USA

Evening visitors to Manhattan's Madison Square Park watch the Oval Lawn pulse with a matrix of ground-hugging light beams, sensor-activated by the heartbeats of a person touching sensors on a metal stand. For this *Pulse Park* performance, kinetic artist Rafael Lozano-Hemmer and his technical team installed 200 Source Four narrow-beam spotlights around the lawn perimeter. These were activated by sensors installed in an on-site sculpture, powered by a portable generator and controlled by custom software on a laptop, linked to a DMX controller and a rack of dimmers. When each participant touched the sculpture's sensors, the loop of lights pulsed sequentially to signal the systolic and diastolic rhythms of their heart. Lozano-Hemmer, a Mexican-Canadian based in Quebec, was inspired for this concept by the minimal music compositions of Conlon Nancarrow, Glenn Branca and Steve Reich, and by a scene in Mexican filmmaker Roberto Gavaldón's 1960 film *Macario*, where the protagonist hallucinated about all living people as candles alight in a cave.

Technology | 200 x Source Four narrow-beam spotlights, heart rate sensor, computer, dimmer rack, portable generator, bespoke software, DMX network controller.

Environments

Parks | Plazas | Promenades

Site Specifics

Helsingborg Waterfront
ÅF Lighting

Helsingborg, Sweden

Opposite top | Fibre optic light sparks, inspired by stars, and circles of ripples projected across paving are the main illusions supporting a 'sea and sky' concept for Helsingborg's recent waterfront upgrade. Designers ÅF Lighting illuminated new pergolas, and used strategies to avoid glare and light barriers interrupting ocean views.

Technology | 18 x Philips PROflood 3000K 150W metal halide gobo spotlights, 17 x Bico custom gobos, 18 x iGuzzini MaxiWoody 5651 35W spotlights with louvres, 14 x Philips LEDline2 LED luminaires, 2 x Roblon IP44 FL fibre optic light projectors with glimmer wheels, Roblon cables.

Aalborg Harbour
ÅF Lighting

Aalborg, Denmark

Opposite bottom | Custom-designed double-spot streetlamps, some with gobo projection filters, scatter light patterns across Aalborg's newly paved waterfront promenade. Plaza spotlights and artistic 'street optics' are scenographic strategies recommended in ÅF Lighting's 'New Nordic Lighting' urban design manifesto.

Technology | iGuzzini MaxiWoody lamps with custom louvres, Philips Pompei IP67 MBF 505WB uplights, Philips PROflood gobo projectors on poles, Philips DecoScene DBP523 uplights, Santa & Cole Àrea light bollards, Simes MEGAEOS square luminaires, ERCO Axis LED walklights, fibre optics, and metal halide, fluorescent and LED lamps.

Ishøj Station Plaza
ÅF Lighting

Copenhagen, Denmark

This page | Flows of water inspired the art concepts for a recent townscape upgrade around Copenhagen's Ishøj train station. As well as functional lighting from LED streetlamps, bollards and recessed wall lights, the project team used marine greens and blues for LED uplighting of trees, and designed lamp beam filters (gobos) to project a luminous scenography of multi-coloured ripple effects on horizontal and vertical masonry.

Technology | Louis Poulsen 70W 3000K metal halide lamp, BEGA 2294 downlights, iGuzzini Linealuce fluorescent lamps, ERCO Lightmark floor washlights and bollards, Philips DecoScene uplights, Philips PROflood gobo projectors, gobo filters.

Liquid Eruptions

El Circuito Mágico del Agua
(Magic Water Circuit)
Various designers

Lima, Peru

Thirteen spectacularly illuminated fountains draw evening crowds to the Magic Water Circuit at Lima's Parque de la Reserva. Opened in 2007 after major renovations to the city's 1920s Parque de la Exposición, the fountains are each designed to represent a different conceptual theme, with computer-controlled water 'dances' synchronized with colour-changing LED light routines.

This page top | *Fuente del Arco Iris* (Rainbow Fountain).
This page centre and bottom | *Fuente de la Fantasía* (Fantasy Fountain).
Opposite page | *Fuente Túnel de las Sorpresas* (Tunnel of Surprises Fountain).

Technology | All fountains: 2,655 x RGB LED lamps, 2,852 x water jets, 288 x water pumps.

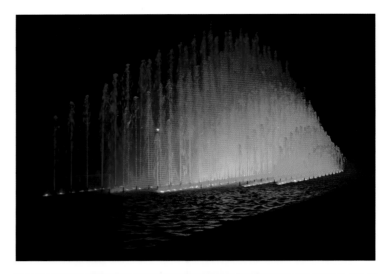

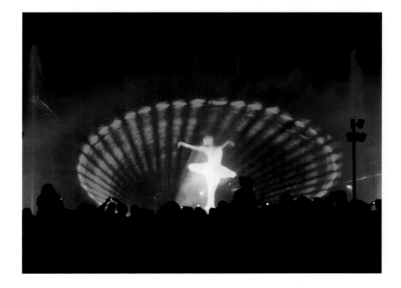

Environments   **Parks | Plazas | Promenades**

140

Pillars of Society

Mingus Streetlamps
Atelier H. Audibert

Le Havre, France

Top | To enhance night uses of central Le Havre, the UNESCO World Heritage-listed French coastal city, Paris-based lighting designers Atelier H. Audibert created new streetlamps. The 3000K LED luminaires were developed by Santa & Cole and Inédit Lighting. Produced in 4, 6 and 8m heights (13, 20, 26 ft), they include three independently controlled light sources that cast different qualities and effects of white radiance. The top lamp, giving a dazzle effect, incorporates hundreds of dimmable LEDs with lenses directing light to the ground. The central shaft comprises four LED strips wrapped with micro-perforated metal sheets to cast subtle moiré patterns. Each base includes two small spotlights to highlight nearby landscape features.

Technology | 276 x 3000K white LED luminaires for footpaths and roads.

Firalet Square
artec3 Studio, RCR Architects

Olot, Spain

Opposite bottom | Fifty kilometres (30 miles) from Girona on Spain's Costa Brava, the history-rich inland town of Olot recently upgraded Firalet Square (also known as Bishop Guillamet's Walk) with a loosely spaced glade of glare-free white LED lightpoles, designed by artec3 Studio and RCR Arquitectes. There are two models: one around the square edges, using 150W ceramic metal halide lamps with road optics; the other near trees in the central area, using RGB 25W LEDs.

Technology | Two types of bespoke lightpoles with white LEDs and metal halide lamps, with diffuser cover plates integrated into the poles.

Friendly Airspace

The High Line
L'Observatoire International

New York, USA

Like a traditional Oriental scroll garden with special places gradually revealed to wanderers, New York's aerial, lineal park, the High Line, causes people to linger and appreciate its lighting and landscaping subtleties. At night, the park's low-height white LED downlighting system, designed by L'Observatoire International, avoids glare, creates a dream-like experience of walking the slatted timber pathway, and enhances outlooks to brightly lit buildings and streets around the western side of downtown Manhattan. The High Line, built in 1934 as an elevated freight line to transport carcasses to the Meatpacking District, was closed in 1980 then redeveloped as public open space by the City of New York in the 2000s, after lobbying by local preservationists, Friends of the High Line. Landscaping, designed by James Corner Field Operations with plant expert Piet Oudolf, includes dense borders of self-seeding grasses, perennials, hardy shrubs and native flowers, inspired by the type of 'weed' vegetation that naturally grew around the old train lines. Architects Diller Scofidio + Renfro designed outdoor furniture using reclaimed hardwood from the railway sleepers, rusty steel rails and blocks of concrete.

Technology | Various custom-designed LED luminaires installed along paths and under seats, steps and handrails, including bollards and some fluorescent tubes mounted on ceilings of passages through buildings.

Environments

Parks | Plazas | Promenades

244

In Absentia

Reflecting Absence:
National September 11 Memorial
Fisher Marantz Stone, Michael Arad,
Peter Walker & Partners

New York, USA

Inside the square pits formerly occupied by the twin towers of New York's World Trade Center, Manhattan's cacophony is quelled by waterfalls tumbling down the granite walls. Lighting designers Fisher Marantz Stone installed LED bars around the bases of these walls, to up-wash the waterfalls with white light, as a glowing tribute to the humans lost and to symbolize hope. Across the ground level parkland, they installed custom-designed LED lightpoles (with T8 rods diffused by prismatic reflectors). As well as the permanent *Reflecting Absence* installation, the Municipal Art Society of New York annually presents 'Tribute in Light', a one-night, bio-fuelled display of twin light beams projected 240km (4 miles) into the sky from 88 Space Cannon 8000W xenon lamps placed in two square formats near the memorial site.

Technology | Waterproof 24V LED luminaires, bespoke lightpoles each using 4 x T8 LED tubes and prismatic reflectors.

146

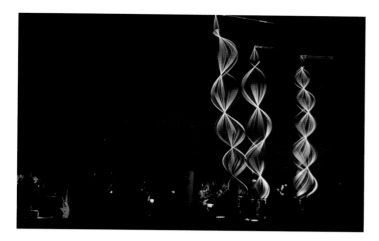

Architectures of Flux

1.26
Janet Echelman

Singapore and other locations

Opposite page | Inspired by traditional net-fishing on the coasts of southern India, American artist Janet Echelman has developed sophisticated aerial sculptures, intricately knotted from high-tenacity polyester fibre and animated at night with multi-coloured projected lighting. In 2010, she created her internationally exhibited *1.26* work for the Biennial of the Americas at the Denver Art Museum, responding to a NASA discovery that the Chile earthquake that year may have slightly redistributed the Earth's mass, shortening the length of the day by 1.26 microseconds. The sculpture's 3D form was generated via amplitude modelling of the Pacific Ocean tsunami triggered by the Chile earthquake, using data from NASA and NOAA. The net is woven with Honeywell Spectra fibre cords, which are extremely strong and light, and (unlike steel) can ripple in response to atmospheric changes. Shown here at the 2014 'iLight Marina Bay' festival, the sculpture measures 24.4 x 18.3 x 9m (80 x 60 x 29½ ft).

Technology | 17 x Martin Professional Exterior 400 RGB LED image projectors, 21 x Martin Exterior 410 RGB LED image projectors.

Sunken Garden
Paul Friedlander

London, UK, and other locations

This page | English kinetic light sculptor and scientific artist Paul Friedlander creates ephemeral, magical 'wave sculptures' with a unique 'chromastrobic' lighting technique. Using modified LED stage lights controlled by customized electronics and self-written software, he causes light to change colour faster than the human eye can see. Yet the colours are revealed when they illuminate fast-moving objects, such as these columns of rapidly vibrating strings shown at Kensington Palace Gardens in 2014. Educated as both a scientist and fine artist, Friedlander has exhibited in many countries.

Technology | 9 x LED stage lights, each with 54 x 3W Cree LEDs and narrow beam lenses (internal electronics modified), DMX interface replaced by a custom PWM interface, bespoke control software.

Fluid Dynamics

Into the Blu
Sophie Guyot

Ljubljana, Slovenia, and other locations

Top | Eighty luminous 'vegetable blooms' in sea blue, green and turquoise have been sprouting up at European festivals since 2009. Delivered by Swiss artist Sophie Guyot, these 'space gardens' are accompanied by sounds muffled inside wooden boxes and transmitted via miniature speakers. The 'flowers' are white Lycra shades stretched on metal frames atop 190cm-high (75 in) curved poles.

Technology | LEDs powered with recycled batteries and coloured with gel filters, white Lycra lampshades, metal frames, sound devices, wooden boxes, speakers.

Lines up. a recollection
Jeongmoon Choi

Berlin, Germany, and other locations

Opposite bottom left | In the courtyard behind a building on Linienstraße in Berlin, Korean-born artist Jeongmoon Choi 'drew' an intricate aerial pyramid with strands of fine cord fixed to the buildings at corners and apex. This delicate construction appears phosphorescent white, like a spider's web, under ultra-violet (black) light. This 2012 installation, one of Choi's 'Drawing in Space' series of ephemeral works in international venues, measured 11 x 11 x 7m (36 x 36 x 23 ft).

Technology | Ultra-violet lighting, cords.

The Pool
Jen Lewin Studio

Black Rock City, USA, and other locations

Opposite bottom right | First shown at the 2008 'Burning Man' festival in Nevada, Jen Lewin's *The Pool* (and its 2014 update, *The Super Pool*) has been delighting impromptu performers all around the world. Participants activate lighting effects by stepping on circular footpads which contain sensors, a wireless radio and LEDs. These change the colours of Lewin's circles to form complex, unpredictable visual effects. Each pad responds uniquely to weights, speeds and areas of pressure.

Technology | 100-120 x circular polyethylene footpads, microcontrollers, wireless Xbee radio, sensor array and 180 pixel RGB LEDs with full colour-mixing and fading controls, customized control system.

Plastique Fantasies

My Public Garden
TILT

Singapore

Opposite top | French light artists TILT (founded in 2001 by François Fouilhé and Jean-Baptiste Laude) create fantastic gardens of luminous 'plants' and radiantly colourful 'flowers' at city festivals around the world. Their 'nightscapes' are assembled with fifteen different light 'bouquets', like frozen fireworks explosions, ranging from 3.8 to 12m tall (12½–39 ft) and weighing from 200 to 2,000kg (440–4,400 lb).

Technology | 15 x designs of steel-framed outdoor sculptures, using various types of lighting (including LEDs, fluorescent tubes and diodes) for 'flowers'.

La Fontaine aux Poissons (Fish Fountain)
BIBI

Lyon, France

Opposite bottom | French artist BIBI (Fabrice Cahoreau) creates fantastic LED-illuminated follies and creatures from orange plastic traffic cones and polyethylene liquid storage containers. For the 'Fête des lumières' in Lyon in winter 2008, he showered radiant, multi-coloured, tropical-style 'fish' around the 1856 stone fountain at the Place des Jacobins in Lyon's UNESCO World Heritage zone. The composition (suspended from a crane) was inspired by underwater vistas seen by the artist while diving. The fish were illuminated with colour-changing LED bars, projectors and stroboscopes to cast dynamic shadow patterns onto the fountain.

Technology | 24 x Magnum Varialed LED bars, 4 x light projectors, 4 x stroboscopes, 4 x Philips VARI*LITE spotlights, DMX controls, 24 x plastic fish.

BIBI's Hell, It is Here
BIBI

Geneva, Switzerland

This page | A plague of 250 tiny, demonic creatures – elves, goblins and banshees – infests a leafless winter tree on the corner of All Souls and Purgatory streets in Geneva. These radiant horror figures were created by BIBI and internally lit with LEDs.

Technology | 125 x low-energy bulbs and 125 x LEDs, 10 x waterproof fairy lights, 250 x devil dolls handmade from plastic bottles.

Public Enlightenment

Darling Quarter Children's Playground
Lend Lease, Speirs + Major, ASPECT Studios

Sydney, Australia

This page | With many water features, swings, slides, a flying fox, sandpits, climbing nets and other engaging structures, the 1.5 hectare (160,000 ft²) children's playground at Sydney's Darling Quarter is magnetic for its target audience of families. Designed by ASPECT Studios landscape architects, it is magically illuminated in the evenings. The lighting, by Lend Lease with London-based Speirs + Major, uses different light sources to provide a vista of theatre and intrigue.

Technology | Blue LED wayfinding lights along dwarf wall, 3000K post-mounted LED down-spots for pathways and features, LED up-spots mounted on timber posts topped with blue LEDs for palm trees and kiosk ceiling structure, gobo projector luminaires for moving patterns across water play area, LED bud lights in grow walls beside kiosks.

Luminous at Darling Quarter
Ramus Illumination, Lend Lease

Sydney, Australia

Opposite page | Like a colour-changing stage backdrop for any big concert event, the *Luminous at Darling Quarter* interactive light wall provides a crowd-riveting spectacle. Visitors can use a mobile app, a local touchscreen kiosk or a website to 'paint' light effects and video games on this façade in real time, or watch some of the 180 hours of pre-programmed content that plays across the building six nights a week. Designed by Lend Lease with content from concert lighting expert Bruce Ramus, this 150m-long (490 ft) elevation of a new Commonwealth Bank building is entirely powered by solar collectors on the building roof.

Technology | 553 x Klik Systems RGB LED linear fixtures, Coolux media server, Fingermark Winsonic touchscreen kiosks with Nexcom NDi5 PC control.

Environments

Parks | Plazas | Promenades

Dancing Pavements

Ice House Square
Studio Fink

Swansea, Wales

This page | Colour-changing and variegated stripes of light streak diagonally across the orthogonally gridded stone paving at Ice House Square, a new waterside plaza at Swansea's regenerating docklands. Designed by British light artist Peter Fink, these Philips Color Kinetics RGB LED strips are inlaid flush with dark granite grid paving. The LED strips delineate light 'rooms' activated by motion sensors in response to pedestrians using the square.

Technology | Philips Color Kinetics LED node lighting, Pharos control system.

BruumRuum!
artec3 Studio and David Torrents

Barcelona, Spain

Opposite page | Barcelona's Plaça de les Glòries, near the iconic and colour-changing Agbar Tower and Design Museum, is a voice-activated playground of more than 500 RGB LED strip consoles, installed at oblique angles across the orthogonal grid of masonry paving squares. A large pole-mounted trumpet attracts visitors to shout sounds such as 'bruum ruum!' to trigger the plaza's array of sensors and temporarily change the colours and patterns of ground-lighting. Around the square's edges, white LED downlights have been installed to guide the paths of pedestrians and help strengthen perceptions of darker skies at night.

Technology | 522 x Instalight 1060 RGB in ground LED luminaires with DMX control system using 9,396 channels, Sennheiser MKE2-P condensor clip-on microphone and ME66 mnicrophone head ambient sensor, Madrix software.

Environments

Parks | Plazas | Promenades

Streets, Stairs, Bridges

In physical terms, cities are complex 3D compositions of structures, spaces and connections. Artificial lighting must be carefully planned to help people appreciate and navigate these conditions at night.

Specific lighting techniques are needed for linear traffic conduits such as roads, bridges, stairways, monorails, promenades, motorways, and even (as in Venice) boat-busy canals and lagoons.

The repertoire of lighting options for wayfaring includes smartpoles, many kinds of streetlights, bollards, under-lit handrails and benches, catenary (necklace-style) strings of lamps between poles, wall-recessed downlights (common for steps), sconces, illuminated objects, digital signs or screens, fairy lights for tree foliage, beacons atop bridge pylons and other landmarks, LED strips highlighting gateways, road portals and bridge cables, and projectors using filters to cast patterns across pavements.

Increasing numbers of wayside fixtures are solar-powered, with photocells and other sensors to switch lamps on and off at preset times, or according to surrounding light levels or human movements.

Electrical engineers are educated to select and modify lighting equipment by balancing a cluster of critical performance factors. Some project budgets are too small to custom-analyze product behaviour in detail, but, on major civic infrastructure projects, experts carefully compare:

— Illuminance: the amount of light affecting a surface, expressed mostly in lux (number of lumens per square metre).
— Luminance: the brightness of a surface quantified in candelas per square metre.
— Colour rendering: the ability of a light source to display colour accurately, on a theoretical CRI (colour rendering index) scale of 1 to 100 (100 = daylight).
— Colour temperature: with warm (reddish) light below 3500 Kelvin (the temperature of white light) and cool (bluish) light above 3500K.

— Direction and beam distribution: including heights of luminaires, shadow effects, positioning and distances between fixtures, beam directions, diameters and diffusions, and 'spills' of extraneous light.
— Durability and other design aspects of the fixtures.
— Energy efficacy: ratio of light output to energy consumption, expressed in lumens per watt.
— Energy efficiency: amount of emitted light divided by the amount of light from the source (a percentage).

These and other performance factors can be realistically visualized, using parametric or procedural modelling software and calculation engines crunching spreadsheet data. Lighting design and engineering programs can virtually test expensive equipment before actual products are installed.

In our nascent millennium of remote sensing, satellite imaging, robotics, machine learning and algorithmic data analytics, there are many promising software-enabled convergences between physical and virtual environments.

One example is the serious games movement, applying video graphics, programming protocols and player strategies to generate new solutions to real-world challenges. Another is the trend towards urban interaction scenarios, where users of mobile devices wander around a city, using augmented reality apps (QR scanning, for example) to access information about specific sites viewed in their portable lens. Services such as Twitter, Facebook and Instagram allow notes and images to be instantly uploaded to inform other users online.

Cellphone-carrying citizens are constantly trackable via GPS signals, and their routes can be mapped on public screens.

History's first phone data mapping demonstration was the MIT SENSEable City Lab's *Real Time Rome* exhibit at the 2006 Venice Architecture Biennale, followed by Joanne Jakovich and Jason McDermott's *Smart Light Fields* live online data map at the 'Vivid Sydney' festival in 2009.

Traffic simulations – from crowds forcing the gates at major events to chaotic driver behaviours in Indian cities without effective road rules – include some of the most complex reality representations emerging in the 'new science of cities' movement (promoted most prominently by Michael Batty from University College London and Stephen Wolfram). Transport logistics modellers are prompt in adapting advanced engineering simulation systems from the aerospace and defence industries.

The Netherlands claims the world's first motorway installation of LED streetlamps, using Philips's SpeedStar models, with sensor equipment, on its A44 highway. These luminaires create a clean white light for improved driver visibility and can be dimmed from 100% intensity during peak evening hours to 20% during low-traffic night hours.

During the next decades, traffic modelling will be integrated with data-rich satellite visuals of built and natural environments, and the discipline-specific modelling systems and datasets used by engineers, architects and designers. Time is being added to many computer models of urban areas, so video visuals more often include lighting effects at night.

What are the art potentials in the safety-scrupulous realm of government-specified transit lighting? Elaborations are the answer. Bridges, especially, offer opportunities for luminous spectacles that emphasize their impressive silhouettes and structures.

158

Ethical Invasion

Packaged River
Luz Interruptus

Caracas, Venezuela

When rain storms across the Venezuelan capital, Caracas, its Guaire River inconveniently spills into certain streets. This phenomenon inspired Luz Interruptus, Madrid's provocative team of street artists, to 'repackage' the overflow to remind citizens of the positive significance of water. Invited by the Spanish Embassy to create an artwork for the annual 'In the Middle of the Street' festival in the Chacao neighbourhood of Caracas in September 2012, the Luz team 'flooded' a narrow street with 2,000 transparent plastic bags filled with clear water. Each was 'a small, ephemeral aquarium', containing a goldfish, pond foliage and a waterproof lithium battery-powered LED unit. For ten hours, onlookers marvelled at the *Packaged River Occupying the Street* – a linear sprawl of small, radiant, water ecosystems available to carry home and decant into a permanent container. During five days of preparation, Luz Interruptus was supported by local groups Bicycle Workshop, Chacao Culture and Green Banana.

Technology | 2,000 x lithium battery-powered LEDs.

Culture Transfusions
Preparations for a Possible Future
Geert Mul

Hagen, Germany

In 2010, Germany's Ruhr manufacturing area became
Europe's Capital of Culture, prompting its neglected towns
to celebrate their histories and focus on new potentials.
Dutch artist Geert Mul led a team which transformed a
motorway bridge through the centre of Hagen. They attached
RGB LED spotlights to twenty lamp posts, then precisely
projected beams of colour-changing light onto tarmac-laid
circular canvas art prints. The changing colours of light altered
the appearances of the artworks remarkably, often creating
simple animations.

Technology | 20 x RGB colour-cycling LED spotlights, 20 x 3m (10 ft) diameter
digital art prints.

162

Step Lightly

Välkommen hem (Welcome Home)
Aleksandra Stratimirović

Stockholm, Sweden

Carving a steep cut into the hill of a new residential precinct at Årstadal, Stockholm, this remarkable public staircase is highlighted at night by colour-changing RGB LED T2 bars integrated with the step risers. Stockholm-based light artist Aleksandra Stratimirović arranged the light tubes in a syncopated rhythm, emphasized with a DMX-timed sequence of subtle, gradual changes in the light colours. Her concept for this work, named *Välkommen hem* (Welcome Home), was an 'unrolled carpet of light'. The 202cm-long (79½ in) LED bars are installed across most of the width of the 13m-high (42½ ft) staircase, which was designed by Nivå Landskapsarkitektur to include an open steel handrail on one side and a solid balustrade of wooden planks on the other.

Technology | Annell T2 opal RGB C/C bars.

164

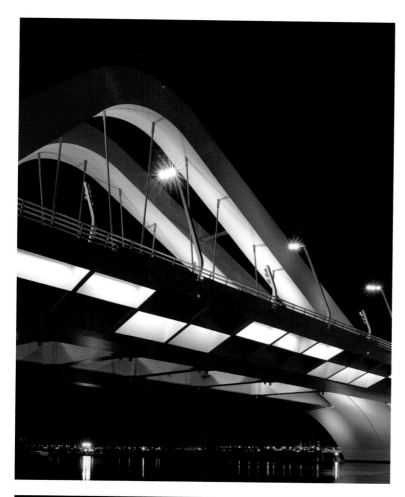

The Helix Bridge / WattFish?
Arup, Cox, Architects 61 / Meinhardt Light Studio

Singapore

Opposite top left | A network of colour-changing LEDs highlights the DNA-inspired double-spiral steel superstructure of Singapore's Helix Bridge. The LEDs are fixed on both spirals to face outwards (highlighting the structure from a distance) and inwards (to the walkway and its canopy).

Opposite top right | For the 2010 'iLight Marina Bay' festival, Meinhardt Light Studio elaborated the Helix with whimsical, 'fishing lines' of clear LED tubes activated by hand-cranked electricity generators. The work – entitled *WattFish?* – alluded to Singapore's origins as a fishing village.

Technology | Opposite top left: Bespoke RGB SMD LED stainless steel luminaires, e:cue power and control system, Bentley BIM software. Opposite top right: Hand-crank electricity generators, LED tubes.

Kurilpa Bridge
Arup, Cox Rayner, Baulderstone

Brisbane, Australia

Opposite bottom | Kurilpa is the largest tensegrity bridge ever constructed. The design relies on a network of steel struts that never touch, but are held in continuous tension by pre-stressed cables and connectors. The superstructure of masts and cables is lit by 144 LED luminaires, mainly solar-powered.

Technology | 84 x photovoltaic panels, Xulux Space Cannon RGB/DMX LED luminaires: 46 x Zeus 6°, 32 x Helyos 20°, 8 x Helyos 30°, 32 x Helyos rectangular and 32 x Nike 9, Xulux DMX/ethernet control system.

Sheikh Zayed Bridge
Arup Lighting, Zaha Hadid Architects

Abu Dhabi, UAE

This page | Irregular waves of massive concrete undulate across the Al Maqta waterway between Abu Dhabi and Dubai. This exceptional pylon system is spectacularly lit with thirteen performance sequences for different seasons and events.

Technology | Bespoke road light masts including Philips 400W and 250W metal halide asymmetric washlights, feature light masts including Martin Exterior 600 575W metal halide colour beam lights, Martin Architectural 55W fluorescent tubes, Martin Exterior 600s with 575W metal halide lamps, Art-Net control network, Martin Pro Maxxyz control software.

Dappled Ground

Broken Light
Daglicht & Vorm

Rotterdam, The Netherlands

Naturalistic lighting, reminiscent of a forest cathedral, has converted Rotterdam's formerly seedy Atjehstraat, in the old Katendrecht docks district, into a delightful environment for today's gentler residents to take evening strolls. Dutch designers Daglicht & Vorm (Rudolf Teunissen and Marinus van der Voorden) developed a chiaroscuro concept involving 'columns' of white light at intervals along the brick building façades and custom-designed streetlights casting calculated refracted light and shadow patterns of foliage across footpaths. The design team prevented glare from entering the windows of street residents.

Technology | Custom-made glare-free LED streetlights.

Environments

Streets | Stairs | Bridges

Street Life

Salon / Sweet Home
Aleksandra Stratimirović

Belgrade, Serbia, and Singapore

This page top | Belgrade contains neglected areas that are architecturally ruined and heavily polluted. For the 2011 'Belgrade of Light' festival, Stockholm artist Aleksandra Stratimirović imagined transforming one degraded city street into a more humane place. With red- and white-striped emergency barrier tape, she crafted a collection of oversized lampshades suspended across the road and a parking zone on aerial cables. Originally titled *Salon*, the work was renamed *Sweet Home* for display in a carpark during Singapore's 2012 'iLight Marina Bay' festival.

Technology | 20-25 x Philips D17 75W LED bulbs.

Cumulus
Ruth McDermott, Ben Baxter

Sydney, Australia

This page bottom | Wild storms and spectacular sunsets inspired the moody lighting dynamics transforming a cumulus-style cloud of perforated and bracket-joined aluminium sheets suspended above a sandstone staircase in Sydney's historic The Rocks precinct. A 2012 'Vivid' festival installation by lighting designers Ruth McDermott and Ben Baxter, the *Cumulus* structure was activated using a Xenian LMX system.

Technology | 50 x Xenian LMX LEDs and iPlayer control system.

Hopscotch
Meinhardt Light Studio

Sydney, Australia

Opposite page | Sensor-activated beams of coloured light blaze in dynamic rainbow stripes across historic stone steps near the Sydney Opera House. Designed for the 2011 'Vivid' light festival by Fiona Venn and Reinhard Germer of the Meinhardt Light Studio, this interactive work used colour-changing RGBW LED spotlights installed into handrails at both ends of each step tread.

Technology | 46 x Klik Systems LEDPOD 50 RGBW custom interactive LED spotlights, Omron E3F2-R4B4 photo-electric switch sensors.

Environments

Streets | Stairs | Bridges

170

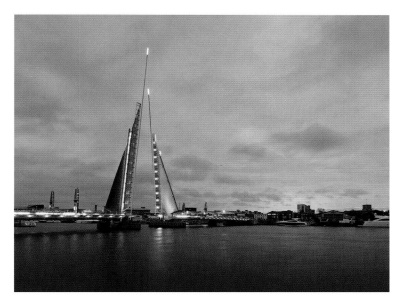

Night Crossings

Mersey Wave
Studio Fink

Liverpool, UK

Opposite page | Liverpool's *Mersey Wave*, by light artist Peter Fink, is a dramatic gateway for the city's drivers and pedestrians. Sited near the Jaguar car plant, it comprises twelve fins fixed at different angles from vertical: the two central fins are 30m (98 ft) high.

Technology | 120 x acdc Taurus blue LED light points with polycarbonate diffusing domes, 18 x acdc Artemis uplighters.

Twin Sails Bridge
Speirs + Major, Wilkinson Eyre

Poole, UK

This page top | An animated lighting sequence engages users of the new Twin Sails Bridge at Poole, on England's Dorset coast. Designed by London architects Wilkinson Eyre with lighting by Speirs + Major, the bridge glows an elegant warm white in its horizontal form. When marine traffic needs to pass, a mechanical operating system separates the bridge along a diagonal joint in its carriageway. As the two sail-like triangular leaves are lifted, a gradual wave of red light rolls down their outer edges, eventually engulfing any waiting pedestrians. Remaining parts of the 'sails', and the tops of two carbon fibre masts, are lit in white.

Technology | Crescent Lighting linear fluorescents, Metamont/Abstract AVR LEDs, iGuzzini LEDs, Mike Stoane Lighting metal halides, Encapsulite linear fluorescents, Sill metal halides, control system with time clock and photocell.

LightBridge
Susanne Seitinger

Cambridge–Boston, USA

This page bottom | To celebrate MIT's 150th anniversary in 2011, smart lighting expert Susanne Seitinger installed a low-resolution 'urban screen' along the Harvard Bridge. Custom acrylic LED diffuser tubes were mounted on the balustrade to create a softly glowing line of 3 x 2,400 pixels – a rare aspect ratio. Four hundred sensors, detecting pedestrian movements, activated different colours and light patterns, reflected across the Charles River.

Technology | 400 x Panasonic PIR proximity sensors, 192 x Philips iColor Flex 50 node SLX LED strings, 96 x Philips Color Kinetics PDS60 power and data supplies, 100 x Arduino Mini control boards, 10 x Arduino UNO control boards and Ethernet shields, 17 x Cisco network switches, diesel generator.

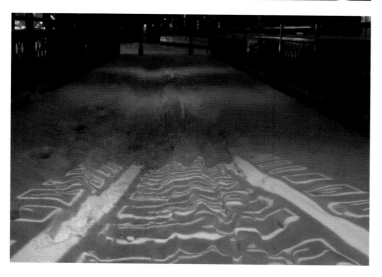

Ambient Passages

Bridge at Cherry Orchard Cemetery
CMA Lighting Design, Fieldoffice Architects

Yilan, Taiwan

Opposite page | Expressing the 'curvature of life' and creating a safe path for evening visitors to a hillside cemetery, this rhythmically spaced sequence of LED floor lamps uplights the rough concrete wall of a curved 'flyover' bridge. The spacing of the lights infers either a faster or slower passing of time, the perspective changing with viewer positions. The scheme was created by Ta-Wei Lin (CMA) with Huang Sheng-Yuan (Fieldoffice Architects).

Technology | 20 x GL Lighting 2W LED uplights, 2 x GL Lighting HPS 70W floods.

Appears@Amsterdam
Titia Ex

Amsterdam, The Netherlands

This page top and centre | Amsterdam's 49m-long (160 ft) Magere Brug (Skinny Bridge) has a central section that can pivot apart to allow tall boats to pass. For this 2012 performance, Titia Ex curved a large sheet of mirror-finish steel to optically contort the bridge so it appeared to warp and dissolve elastically during the daytime. After twilight, it seemed to merge with its surroundings. Clamping another mirror near the water caused the bridge's reflections to fracture like a kaleidoscope.

Technology | Mirror panels fixed with clamps covering bridge.

Current 3 (cubed)
Virginia Folkestad, 186 Lighting Design Group

Denver, USA

This page bottom | Colour-changing light washes and gobo-projected cultural imagery transform Denver's vintage Wynkoop Street Railway Bridge with 'breathing' colours and luminous scenes at night, with some 'events' sensor-triggered by pedestrian movements. This 'kinetic sculpture' was created by local artist Virginia Folkestad with the 186 Lighting Design Group, and was organized by Diane Huntress leading the LoDo public-private heritage art projects partnership with Denver government and neighbourhood agencies.

Technology | 3 x Philips Color Kinetics ColorBlast Powercore units with AuxBox, Data Enabler, iPlayer 3 and ColorPlay control system, 3 x Raztech ODiLite pattern projectors, TECO programmable logic relay.

Retail Wonderland

Regent Street Relight
Studio-29

London, UK

Originally designed as a grand boulevard by Regency architect John Nash in 1811, London's Regent Street is now elegantly illuminated at night with a chiaroscuro scheme of warm and cool white highlights and shadow effects. Since 1995, owner The Crown Estates has gradually relit the entire street from Regent's Park to The Mall, advised by Tony Rimmer of Studio-29 Lighting Design. After extensive negotiations with tenants and government agencies, the team has developed a consistent strategy and equipment palette, mainly using Meyer low-energy and LED spots, floods and projectors, with cold cathode tubes and LED strips. These are customized to highlight notable architectural features on key buildings – mostly constructed from the 1890s to the 1920s as replacements for Nash's modest mid-19th-century structures.

Technology | 3,500+ Meyer low-energy and LED luminaires (various), fibre optics, 2km (1.2 miles) cold cathode tubes, acdc, iGuzzini and Osram LEDs.

Environments

Streets | Stairs | Bridges

Lunar Life in Humane Places

Mary-Anne Kyriakou

Glare from pre-LED streetlights around an intersection.

Dazzling our eyes to compete for valuable public attention across the night spaces of cities, illuminated buildings and billboards offer transitory optical stimuli to arouse the psyches of pedestrians and motorists remotely.

Deeper impressions can be kindled when people experience works of art more closely. Then they can engage directly with the intentions of creators and express personal reactions to companions sharing the scene.

Lighting dark city spaces is an exciting new realm of imagination for creative professionals. Apart from a city's skyscape – its elevated backdrop of privately owned towers and imposing edifices – there are many potentials for transforming dark zones of pedestrian ground: laneways, pathways, plazas, public gardens, ponds, promenades, playgrounds, sports courts, undercrofts, tunnels, bridges and other conduits and crannies.

Whether formally designed for civic ceremonies or accidental and interstitial zones of uncertainty and danger, ground-level urban spaces can be reinvented to provide magical new experiences for every unique user.

In today's third age of light (the electroluminescent era following history's natural-radiant and electric periods), LED lamps and digital systems are providing unprecedented creative scope, control and safety, with public light art performances at night.

Electroluminescent lamps are the first high-powered light sources that are tolerable to touch. They generate different atmospheric effects from hot (and sometimes dangerous) outdoor light sources, such as bonfires, fireworks, fuel-burning lamps, and electric searchlights or lasers strafing the sky.

Local government leaders are now delivering not just multi-functional smartpoles along main streets, and permanent precinct lighting aligned with international 'green city' protocols and renewable energy targets, but also computer-programmable infrastructures, and increasing numbers of 'luminale' festivals, to help light artists activate busy pedestrian zones at night.

Night artists are like fairground sorcerers demonstrating new tricks with electricity during the final decades of Europe's gaslight era. Testing advanced equipment and techniques, they cast light to quicken heartbeats and kindle lingering perceptions.

How humans gaze

Human processes of seeing, recognizing and reacting evolved to help us make sense of situations. Gaze is influenced by our spatial awareness of local circumstances and our emotional responses. Perceptions of environmental characteristics such as openness, closeness, comfort, safety and uncertainty are all affected by the colours, intensities and directions of light beams playing on surfaces. Sounds, smells and air qualities enhance these understandings and influence our behaviours.

In urban environments, artificial lighting and atmospheric conditions can manipulate our conscious and sub-conscious minds, especially at night. Between dawn and dusk, humans have always relied on the sun to illuminate outdoor environments: its radiance casts different qualities of light and shadow on landmarks across fields of ground and water. After dark, vision in cities does not come from the moon and stars but from streetlamps, shop windows, vehicle headlights, and arrays of lights inside or mounted on buildings.

Pedestrian experiences tend to happen at an easy walking pace of 5km (3 miles) per hour. Generally, our binocular vision allows us to scan our surroundings across 180–270° forward, 60° upward and 75° downward. In the city at night, this visual scope is often constrained by rows of streetlights fixed at heights between 7 and 12 metres (25–40 ft) and spaced between 10 and 30 metres (35–100 ft) apart. Illumination from shops, buildings and other city assets fills in the spaces between streetlights, creating backdrops of light, like scenery around a theatre stage.

High illumination levels and glare from streetlights can dazzle people, especially when these are contrasted against zero or low illumination levels either near the ground or looking beyond the lighted area. Natural human processes to assess dark surrounds can also cause general feelings of insecurity. Eyes need time to adjust between brightness and darkness – especially for older people.

The eye has two main types of receptors to control vision. Cone receptor cells assist in colour and detail recognition under day or bright light (photopic) conditions. Rod receptors assist peripheral vision and operate during low light (myopic) and dark (scotopic) conditions.

When eyes must adapt from light to dark conditions, the rod receptors become increasingly active to reach maximum awareness of deep blue and green, then (after 5 to 30 minutes) black and white. Adapting to bright light is significantly faster, with the cone receptors used to interpret bright colours.

These optical effects of eyes adapting to darkness are known as the 'Purkinje shift', as they were first recorded by Czech anatomist Jan Evangelista Purkyně in 1819. He noted that, during twilight hours, red geranium flowers seemed to become darker/duller, while their green leaves gradually seemed brighter.

Purkyně's observations are essential for the design of humane lighting for different environments during evening and night hours. For example, at low light levels (when eyes are in myopic or scotopic mode), people find it more difficult to differentiate bright colours, so cool white light is most useful for basic orientation, while colours may be used for additional interest or theatrical effect.

A specific case is the recent switch by some bus operators from traditional yellow to white illuminated destination signs, facilitated by LED lamps and transistor controls. However, care is recommended when using white light for night scenes to avoid over-illumination and to reduce the high contrast that causes discomfort, dazzle and disorientation.

Designing humane lighting for outdoor areas requires knowledge of how certain types of retinal ganglion cells affect non-visual processes of eyes reacting to environments. These ganglion cells are linked to the body's circadian rhythm (internal clock) and production of melatonin. Even partially blind people, without functioning rods and cones, appear to have their body clocks controlled through light stimulating these specific cells (Zaidi, Hull and Peirson et al, 2007). White LEDs rich in high concentrations of blue wavelengths may be especially disruptive to body clocks and sleep patterns.

Power blackout in Tokyo's Ginza red light district after the Fukishima earthquake in 2011.

In the case of forced artificial lighting use, Japan's Fukishima earthquake disaster of 2011 forced power blackouts and major energy consumption reductions across the nation. Dim buildings and streets, suddenly devoid of video screens and late night office lighting, dulled the senses of people wandering through usually exciting commercial zones such as Shibuya in Tokyo. Some citizens changed evening meetings because they had safety concerns about the gloomier conditions (Zarroli, 2011).

City governments are responsible for the infrastructure and public impacts of outdoor lighting at night. Leading discussion on world's best practices is Lighting Urban Community International (LUCI), a network of 65 member governments and 34 light industry organizations, organized by the City of Lyon in central France.

LUCI's urban strategies commission is eager to educate city planners and lighting professionals about the emerging social issues of public lighting. With the University of Liège's Architectural Methodological Research Laboratory (LEMA), it has published *The Social Dimensions of Light*, a report highlighting thirteen international city lighting success stories. Most of the highlighted projects were from cities in northern Europe and Scandinavia, where some citizens are depressed during winter due to seasonal affective disorder (SAD). More than twenty European city governments have introduced annual winter light festivals in order to cheer their citizens and attract visitors. Best practice of illuminated cities is featured by the LUCI group and includes the illumination of historical and cultural landmarks and locations for enhancement of prestige, city marketing and public appreciation.

Challenges of light pollution

Ubiquitous artificial light is causing persistent sky pollution across most cities. Public lighting patterns of different cities are well documented in Earth images from satellites. Sky glow caused by metal halide and mercury streetlamps clouds views of stars and regularly interferes with bird migrations.

Astronomers and animal activists are supporting the dark sky movement, which is highlighted by public 'switch-off' events, such as the United States 'National Dark-Sky Week' (founded by Jennifer Barlow in Minnesota in 2003) and the international 'Earth Hour' events (organized by the World Wildlife Fund, with its first event in Sydney in 2008). Campaign leaders support regulations to protect 'dark sky oases' and to regulate public lighting use between hours of low activity.

Common causes of light pollution are unshielded streetlights and powerfully illuminated advertising signs. These waste light by dispersing it upwards to release photons (particles of electromagnetic radiation) that bring about illuminated skies due to cloud light reflectance and obstruct the views of the stars, and/or project undesirably intense illumination inside neighbouring buildings. These problems became evident with the massive construction growth and electrification of Western cities after the Second World War. The situation is now exponentially increasing, with development of megacities across Asia, South America, Africa, Russia and the Middle East.

Light pollution is defined as 'excessive and/or misdirected' light, generating glare, light clutter, light trespass and sky glow. A 2010 study by Stanford University student Jodi Shi claims that (in the United States) commercial organizations produce five times more light pollution than operators of residential or industrial buildings.

Stadium lighting creates sky glow over Portsmouth, UK (photo Dark Sky Association).

Advertising entertainments at London's Royal Vauxhall Gardens, 1848.

Evolution of public lighting systems

'The nature of light during the night can be understood in two streams; firstly expelling darkness and transforming the night to day[,] and another (aspect) focused on creating and manipulating darkness, whether in devotion or spectacle' Craig Koslofsky (2011).

Age of natural radiance

Public lighting began with campfires and flaming torches warming prehistoric humans at night. Early communities of *homo erectus* (dating back more than one million years) burned animal fats, beeswax, timber, charcoal, and fish or vegetable oils for light, warmth and cooking.

In the ten centuries before Christ, citizens of ancient communities learned to craft fire bowls, oil lamps, lanterns and other types of luminaires to protect and enhance their flickering light sources. Wicks (made of rice paper in Asia, or cotton and other woven textiles) were used to moderate fuel burning times. Initially clay was the main material used for lamps, but after the discovery of iron, and as people began to settle in villages and towns, metal lamps were hung from wall brackets and tall lamp posts were fixed in sequences along public roads.

When towns installed streetlights, workers were needed to light the lamps at twilight, tend them through the night and ensure they were safely extinguished; also to escort pedestrians safely across dark zones between well-lit public areas. This system became increasingly prevalent during the Roman Empire and as towns grew across Europe and Asia through the first fifteen centuries after Christ.

In 1417, the Mayor of London announced formal laws for public lighting, requiring property owners to tend candles and oil lamps around their street-facing doors and windows each evening. Paris introduced similar laws more than a century later, in 1524.

From the end of the Dark Ages in the 14th century to the dawn of the Enlightenment in the 18th century, European citizens relied on oil lamps to illuminate their evenings at public hotels, pleasure gardens, theatres and coffee houses, and to light the way for horse-drawn carriages. After-dark spectacles, with fountains of water synchronized to live music, lights and fireworks, were popular at the splendid royal gardens of Versailles during Louis XIV's reign over France (1661–1715) and at Italy's Villa d'Este after its initial building works in the late 16th century.

From the 17th to 19th centuries, English pleasure gardens, for example Vauxhall in London, gradually opened to crowds from all social classes. They became popular evening meeting places for royalty, aristocracy, families, and young men and women seeking liaisons. In their 2011 history of Vauxhall Gardens, David Coke and Alan Borg quote impressions from a German prince witnessing 'strange' illuminations past midnight on his 1827 visit:

During supper, one of the great special effects of Vauxhall was enacted. As night fell a whistle was blown as a signal to a number of servants placed in strategic parts of the garden. Each servant touched a match to pre-installed fuses, and, 'in an instant', over a thousand oil lamps were illuminated, bathing the gardens in a warm light that would have been visible for miles around. In the days before electric light, the effect was sensational, and was a constant attraction at the gardens.

In the early 18th century, Swiss chemist Aimé Argand designed a new type of oil lamp. The larger wick and a cylindrical glass chimney could maintain brighter flames that were also safer to carry. This style of 'hurricane' lamp, still widely used outdoors, was mainly lit with whale oil or kerosene.

Gas was not widely used for public lighting until long after English scientist Stephen Hales extracted a flammable liquid from coal in 1726. After several successful demonstrations on properties owned by English industrialist William Murdoch (aka Murdock) in the 1790s, coal-gas streetlighting was first installed along London's Pall Mall by Frederick Albert Winsor (aka Winzer) in 1807. Five years later, the British Parliament granted a charter to the world's first power utility – the Westminster Gas Light and Coke Company.

Baltimore was the first United States city to install gas lamps, in 1816. Among European capitals, Paris installed its first gaslights in 1820 and Lviv (then part of Austria, now in the Ukraine) introduced kerosene street lamps in 1853. By the end of the 19th century, gas lamps lined the main streets of most large cities across the northern hemisphere.

Gas lamps were manually lit and extinguished with a long rod that could open and close the gas taps. In the early 20th century, some cities installed automatic timers, spring-wound to last a week. These devices would maintain a pilot light during the day, then release a lever at pre-set times to control the mantle and flame.

Although gas was less expensive than oil, it was widely unpopular because of its character, a noxious smell, and the intensity of light (aesthetes noted ghoulish black shadow effects and a 'loss of mystery' in public places). Wire gauzes were soaked with chemicals to try to mask the gas odours.

However, gaslight also facilitated new public delights at night – most obviously at early 19th-century theatres and amusement gardens like London's Vauxhall, and the Tivoli parks in Copenhagen, Ljubljana and other European cities. Gas also generated new creativity with light playing on flows of water. Outstanding examples were the Bartholdi Fountain at the 1876 Philadelphia Exposition and the Coutan fountain at the 1889 Paris Exposition (where electric lighting was also extensively introduced).

Late 19th-century Impressionist painters, including Pierre Bonnard, Camille Pissarro and Vincent van Gogh, captured the romance of gas-lit dining at city cafés and restaurants, for example, in Van Gogh's 1888 *Café Terrace at Night*.

Electric era

During pre-electric centuries, public lighting always came from natural sources of heat-generating radiance. A transitional moment occurred in the late 1700s, when Italian inventor Luigi Galvani demonstrated what he called 'animal electricity' by connecting two different metals in series to each other and to a frog's leg as an electrolyte/conductor.

In 1800, Russian count Alessandro Volta demonstrated electric currents via a battery called the 'voltaic pile', using zinc and silver as the different metals, with brine as the conductor. These were the earliest examples of battery power.

Further electric light demonstrations were conducted in London by Cornish chemist Sir Humphry Davy. He connected two wires to a battery and attached a platinum strip (in 1802) and charcoal strip (in 1809) between the other ends of both wires. In a process called 'arc' electricity, these strips (filaments) glowed with light from the batteries. The charcoal filament facilitated what later became known as 'carbon arc' incandescent lamps.

During later decades of the 19th century, more than twenty engineers produced rudimentary (not commercially feasible) designs for electric carbon arc lamps – notably William Edwards Staite, with William Parsons, between 1834 and 1853 in London.

The first electric streetlights were carbon arc lamps using alternating current between two electrodes. Known as 'Jablotchkoff candles', these were demonstrated in 1876 by Russian telegraph engineer Pavel Yablochkov. Paris won its enduring reputation as the 'City of Lights' by installing many Jablotchkoff candles around the

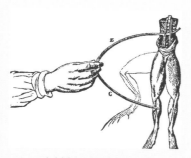

Luigi Galvani proved 'animal electricity' with frogs' legs twitching when connected to both zinc and copper wires.

A carbon arc searchlight used on city landmarks in the late 19th century.

Grand Magasins du Louvre (the world's first department store) in 1878 – the same year that London installed its first electric streetlights around Holborn Viaduct, the Thames Embankment, Billingsgate fish market and Mansion House.

Colonel Rookes Evelyn Bell Crompton demonstrated improved arc lamps, with portable generators of his own design, at the Henley Regatta and Alexandra Palace in 1879. In 1880, he formed a company with Joseph Swan, the first British inventor of electric incandescent lamps, to install and service both types of lighting.

In 1879, American telegraph engineer Thomas Alva Edison also applied to patent an incandescent electric lamp (similar to Swan's), using 'a carbon filament or strip coiled and connected to platina contact wires'. After the patent was granted, Edison successfully tested a carbonized bamboo filament that remained alight for more than 1,200 hours. He won finance for his Edison Electric Light Company to make and sell the lamps, claiming: 'We will make electricity so cheap that only the rich will burn candles.' After pricing and patent battles with US competitors during the 1880s, Edison formed a partnership with Swan to sell incandescent electric bulbs in Britain from around 1889.

In 1881, the world's first 'moonlight tower' (a beacon-topped steel pylon) was built across a major street junction in San Jose, California. According to an article by Kris De Decker for the online journal *Low-Tech Magazine*, its tapered iron engineering design inspired other moonlight towers across the US, and the especially graceful and elaborate Eiffel Tower, which was switched on for the Paris Exposition in 1889.

Occasionally arranged as 'grids' to cover an entire city (for example, 122 towers across Detroit from the 1880s to around 1910), moonlight towers each supported four to six arc lamps (of about 2,000 to 6,000 candlepower) on a platform elevated up to 90m (295 ft) high. They were low maintenance compared to oil lamps and could floodlight large commercial areas, industrial zones, ports and sportsgrounds, but were unpopular because they created dark shadows behind buildings, excessive glare for users of the illuminated areas, often burnt all night all year, and were unflattering to the faces of pedestrians.

Despite safety and pollution concerns, electric lighting, combined with the spread of both automobiles and transport infrastructure, enabled what is now known as 'the night economy' of cities across Europe and the US. Also, Edison's commercial acumen and patented innovations sparked exciting advances by pioneers of the electronic arts. Innovators developed new lighting techniques for public fountains, shop windows, parks, plazas, foyers, billboards and architectural icons, transforming the atmospheres, spectacles and experiences of modern cities at night.

Also defining the dark hours of new electric cities in the late 19th century were various types of public signs: most memorably Hector Guimard's cast-iron gateway portals identifying the first Paris Métropolitain subway stations launched in 1900.

For wealthy and adventurous citizens, electric illuminations fostered new interest in going out at night. Shopping hours were extended in the late 19th century as women and servants began to acquire more legal rights and disposable budgets to spend on clothing and household furnishings. With mechanization in factories and transport, larger sizes of plate glass became available for windows, thereby enlarging merchandise display frontages for shops (Dan and Willmott, 1907).

By the end of the 19th century, heights of commercial buildings in fast-growth US cities (especially New York and Chicago) had soared far beyond the human scale historically known in European and Oriental capitals. Heights of streetlights also increased to expand zones of public illumination – an issue of importance as motor cars increasingly challenged the lives of pedestrians.

Los Angeles cemented its reputation as one of the world's first car-oriented cities with early 1920s development of the Miracle Mile – a 2.4km (1.5 mile) section of farmland road that is now called Wilshire Boulevard. In a 2013 history of the Miracle Mile, Ruth Wallach recorded that entrepreneur A. W. Ross sold roadside sites for large retail stores and shopping malls, insisted on bold building designs and signage to attract drivers behind the windscreens of moving cars, introduced dedicated turning lanes to encourage parking, and installed timed traffic lights to limit pedestrian interruptions of vehicle flows.

By the 1930s, the topography of artificial light in American cities almost always included car-related fixtures such as illuminated pavement bollards and traffic lights. These mediated time and space allocations between cars, pedestrians and (increasingly rare) horses.

In the 1920s and 1930s, theatre designers began to be employed to light the shafts and spires of key skyscrapers in New York and Chicago. In February 1930, the General Electric Company published an industry newsletter, *Architecture of the Night*, to promote emerging strategies for floodlighting, spotlighting, uplighting, downlighting and beacon effects.

After the Second World War, Western cities were extensively redeveloped using mass-produced equipment and standard design concepts that are now explained generically as 'technical' illumination. In retrospect, manufacturers and professionals often delivered bland or harshly unpleasant evening atmospheres across many cities and suburbs.

In the early 1960s, some English and American intellectuals argued for more humane and nature-sensitive approaches to urban planning and development. Three influential books were: *The Image of the City* by Kevin Lynch, 1960; *Silent Spring* by Rachel Carson, 1962 (which especially triggered the environmental movement); and *The Death and Life of Great American Cities* by Jane Jacobs, 1961. Jacobs argued against 'lifeless' modern approaches to town planning and used the analogies of light and fire to explain her preference for dynamic cities – to be enjoyed as 24-hour melting pots of diverse people.

Electroluminescent developments

Jacobs's concepts were ignored for most of the 1960s, but human-sensitive approaches to urban design and public lighting have been improving since the 1970s. Designers gradually began to emphasize lively scenic lighting as well as standard technical illuminations. With continuing evolution of LEDs and other semiconductor-enabled systems, a new digital language of light has been evolving (Ritter, 2006).

The post-1960s boom in international tourism demands that cities brand themselves with unique identities. Urban lighting masterplans have become increasingly sophisticated, incorporating strategies for safety, identity, culture, history, prestige, advertising signage, pollution, pedestrian experiences and entertainment. Most plans are renewed every eight to ten years.

Bright, technical lighting is still needed for many large areas of cities – especially to provide security for pedestrians walking in parks and public gardens. However, new lighting equipment and design techniques provide better aesthetic integration of lighting in nature-sensitive zones. Indirect lighting – for example, wall washing – is often effective to illuminate spaces comfortably and at a human scale. Also LED lamps now have improved renditions of colours, enabling more subtle emotional qualities. Strategies to illuminate rows of trees and uplight selected trees and features are combined to create after-dark scenes. LED technology also enables dimming and creations of different light scenes for use over annual cycles. Lighting can be programmed and sensitive to local environmental factors, including reduction of lighting levels during periods of low human activity.

Since 2000, lighting has been increasingly integrated in street furniture. Examples include LED-illuminated products, such as the 'Positano' exterior benches by Karsten Winkels for Hess and Ross Lovegrove's 'Solar Tree'

Artemide's Solar Tree streetlight, designed by Ross Lovegrove with aluminium branches, powered by Sharp Solar cells, launched in European cities in 2007.

streetlights, tested in Vienna by Zumtobel. Illuminated advertising consoles sometimes incorporate video screens, and custom lights are sometimes installed to extend evening uses of barbecue tables and seats in parks.

By the end of the 2010s, most urban lighting will use LED lamps and digital control systems, expanding possibilities for lighting manipulation and energy savings in public space: for example, lighting can be dimmed during low street occupancy in early morning hours. Government surveillance devices such as cameras and motion detectors are also being integrated into smart streetpoles.

Designing Night Art Experiences

Public art has been evolving to include light artworks, and urban planning now includes lighting masterplans. These changes are encouraging governments and property developers to fund new light artworks instead of day-lit sculptures, wall murals and other forms of static art. However, old issues about public art recur with public light art. What is good quality? What is conceptually and visually appropriate? How will an artwork be paid for and maintained? These questions will be managed increasingly by curators with technical expertise in outdoor lighting and familiarity with city planning.

Curatorial capabilities are also required as the new artistic and technical potentials of light are interpreted in outdoor city contexts (rather than galleries). One topical example is how to arrange appropriate and qualified curation of internationally magnetic city light festivals – which are not like conventional art exhibitions or fairs. Experienced curators of art galleries and museums are eager to 'control the space' of outdoor festivals, yet they are often not technically qualified to understand the engineering or scientific sophistication of apparently simple displays.

Creating night scenes in public zones requires sensitive artistic manipulation of light and darkness. Functions and interpretations of spaces must be designed in terms of forms, atmospheres, spaces, details and connections. Characteristics of light (such as colours, contrasts and degrees of saturation) must be arranged for orientation, narratives over time, subjective perceptions and overall satisfaction, balancing the scale and proportion of components in a scene.

Emotional transitions across cities have always been defined and enriched by subtle plays of light, shadow, colour and darkness. Electronic artists test their talents – and our moods – by exploiting these dichotomies in wondrous ways.

Exhibits

Opposite: *Parmenides 1 (Star Geode)*
by Dev Harlan, first exhibited at the
Christopher Henry Gallery in Los
Angeles in 2011 (details p.268).

Dynamic Objects

Electroluminescent technologies are rebooting the history of sculpture. When activated by electrical semiconductors, LED-lighted objects are cool and safe to touch, vastly expanding artists' potentials beyond last century's kinetic experiments with hot electric lamps.

Traditional sculpture's greatest legacies include tribal totems of carved wood, Egyptian and Grecian stone caryatids, Italy's ideal humans of white marble, bronze commanders at war on their courageous horses, and modernism's anti-figurative 'turd in the plaza' castings or constructions. These legacy formats generally require surgical skills with sharp tools, to manually model the clay, carve the stone or cut sheets of metal.

In the 21st century, sculpture's potentials have been greatly expanded by the complex form-generating powers of 3D computer design programs. These produce files of code to instruct machines exactly to perform 3D printing, laser-cutting, folding, routing, etching, embossing, glueing and fabrication tasks robotically.

Many design-construct processes devised by engineers of spaceships and aircraft (at costs only affordable in the defence budgets of the world's few wealthiest nations) are now widely accessible to even amateur sculptors on small budgets.

As well as today's exciting advances in making sculptures, artists can use numerous types of light to activate their static forms and expand their creativity. LED-supported concepts include illuminating objects (from within or behind) as lanterns, integrating television screens and sound sources, installing arrays of lamps as programmable light pixels, and projecting colours or images from, or onto, the artwork.

Another approach is to assemble mass-produced lamps creatively and plug them into a power source. Triumphant examples are the 'Keyframes' crowds of giant stick men – fluorescent tube figures that have been performing strobic athletic and dance effects at outdoor events around the world (see p. 190). Inspired by 19th-century photographs of athletes caught in various stages of action, the 'Keyframes' displays were invented by French team Groupe LAPS for the 'Fête des lumières' in Lyon in 2011. They use light to take fixed objects into the dynamic domains of choreography, animation, sport and film.

Light has always been critical to perceptions of sculptures, which historically have been naturally illuminated from the sun or moon, or artificially lit by one or more nearby lamps. One of the world's most outstanding classical architectural examples of all three lighting sources for sculptures is the Pantheon in Rome, where natural illumination is transmitted through the oculus of the dome. Today we have exponential capabilities to integrate not only light but also sound, video and internet-transmitted information as dynamic elements to enliven public sculptures memorably.

Compare the different object-activation strategies of two artists who recently showed large, prismatically complex objects made of robotically cut sheet metal. First is New York artist Dev Harlan, whose *Parmenides I* faceted sphere is bland until he begins to project several synchronized video loops of images that have been precisely computer-mapped onto the object's many differently angled faces (see p.185). Second is Rotterdam artist Daan Roosegaarde, whose *Lotus Dome* is an elaborate silver 'flower' assemblage of laser-cut petals of layered Mylar foil (see p.203). Edges of the petals roll open and closed when touched by light that is affected by the number and proximity of viewers. Initially shown in a Gothic cathedral in Lille, France, the work is internally lit like a lantern to cast flickering light and shadow effects on nearby walls. These interactions are accompanied by deep bass music.

Such works exemplify not only a third millennium revolution in sculpture but also a powerful reinvigoration of the 20th-century lumino-kinetic art movement. Mainly based on interior works that can either really or apparently move, pre-LED kinetic sculptures included a subset of outdoor constructions augmented with electric light (most commonly neon and fluorescent tubes).

One of the first practitioners of kinetic art, Russian constructivist Naum Gabo, wrote (with Antoine Pevsner) a radical agenda for art after the Bolshevik revolution. Published in Russian in 1920 and translated into English in 1957 as *The Realistic Manifesto*, this tract included a counter-intuitive call to 'renounce' colour, line and volume. It noted key propositions of relevance to light sculptors today:

Space and time are the only forms on which life is built and hence art must be constructed... We affirm depth as the only pictorial and plastic form of space... We renounce ... the mass as a sculptural element...

Difficult to clarify with printed images are outdoor light artworks that primarily celebrate space and time and could not be called 'sculpture' in any obvious sense. These include artificially illuminated mist, haze, steam and fog 'episodes' (for example, Anthony McCall's aborted column of spiralling steam for Liverpool and Fujiko Nakaya's fog-wrap of Philip Johnson's Glass House in Connecticut). These have precedents dating back to pagan rituals and seem ironically close to being not art(ificial) but natural atmospheric phenomena.

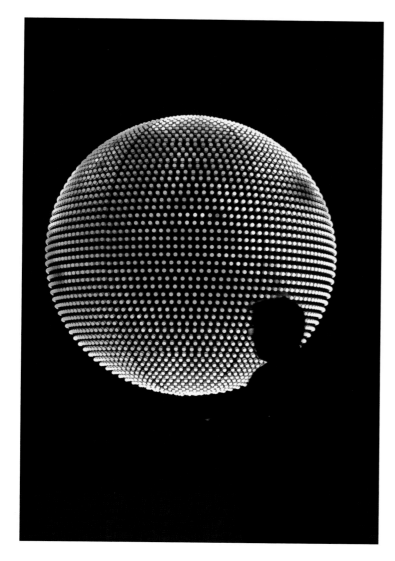

Night Suns

Equación Solar (Solar Equation)
Rafael Lozano-Hemmer

Melbourne, Australia

Opposite page | Animated visuals of the sun's flaming behaviours were projected onto the world's largest aerostat (helium balloon), tethered 20m (65 ft) above crowds attending the 2010 'Light in Winter' festival in Melbourne's Federation Square. Scaled 100 million times smaller than the real sun, this sphere's graphic displays could be altered by nearby users of a mobile app. In a concept by Montreal-based, Mexico-born electronic artist Rafael Lozano-Hemmer, recent NASA images of the sun were overlaid with animations derived from mathematics for fluid dynamics, including reaction diffusion, perlin, fractal flame and Navier-Stoke equations. These were accompanied by computer-generated simulations of rumbles, fire crackles, flares, vents and bursts, as detected by NASA's sun-monitoring instruments. The luminance of the orb has been measured at 110,000 ANSI lumens (reflected projection).

Technology | Airstar Space Lighting aerostat (helium balloon), Antimodular balloon-tracking and 3D projection-mapping software, 5 x high-definition zenithal projectors, mirror, synchronized media servers, DMX network.

The Walk
Titia Ex

Eindhoven, The Netherlands, and London, UK

This page | Medieval writer Dante Alighieri's epic poem *The Divine Comedy* inspired Dutch light artist Titia Ex's 7½-minute video loop dramatically shown 'in the round' on a 2.5m (8 ft) diameter spherical pixel screen. Designed as the centrepiece for the 'Light in Time' exhibition of lighting history at the 2012 'GLOW' festival, Ex's *The Walk* includes images of torch-holding figures slowly circling the sphere towards the afterlife. Viewers are invited to join their eternal procession; to also walk around the sphere in a trance-like rhythm. Ex worked with Philips Color Kinetics and Eindhoven's The Light Art Centre to create the exhibit, which was later shown at the Kinetica Museum in London.

Technology | 5,000 x nodes of Philips Color Kinetics iColor Flex LMX colour-changing RGB LED strings, Pharos LP30 controller, bespoke software and video imagery.

Stop Motion Sports

Keyframes
Groupe LAPS

Lyon, France

Freestyle athletics in urban situations – a non-competitive sport globally popularized from France under the name *parkour* – is being most spectacularly demonstrated by hordes of non-human 'figures of light'. In a travelling outdoor show called 'Keyframes', French light and media artists Groupe LAPS use custom WhiteCAT software and a DMX network to animate several dozen 2m-tall (6½ ft) 3D stick figures by remotely pulsing (on-off) their skeletons of white or yellow LED tubes. Groupe LAPS was formed by Thomas Veyssière and other electronic artists in 2008 and first demonstrated their 'Keyframes' athletes at the 'Fête des lumières' in Lyon in 2011. Although the figures remain immobile, the light pulsations cause observers to perceive them dancing energetically across roofs, building façades, walls and other dangerous structures.

Technology | WhiteCAT, Blender, Live Ableton and Audio Fire (customized) software, numerous double 24V/4000K LED tubes and circles, LED pwm gradation (dimming), DMX network.

Dynamic Objects

Mythic Geometries

Digital Origami Tigers
LAVA

Sydney, Australia, Singapore and other locations

Opposite page | Like imperious stone lions guarding important entrances of classical architecture, two luminous orange tigers crouched symmetrically outside Sydney's colonial Customs House during Chinese New Year celebrations in February 2010. Launching the Year of the Tiger, these 2.5 x 7m (8 x 23 ft) creatures glowed like traditional Chinese lanterns and were shaped in planar forms emulating ancient Chinese *zhezhi* paper-folding techniques. International architects LAVA used advanced computer modelling and laser-cutting systems to build the creatures with aluminium frames clad with stretchy Barrisol (recyclable) fabric, fixing Philips Color Kinetics red LED strings and floods inside each structure. The tigers have been globetrotting to festivals in Kuala Lumpur, Singapore, Berlin, San Francisco and Amsterdam.

Technology | Each tiger: 18 x Philips Color Kinetics IP67 LED strings with 30 red nodes and shared power, 2 x Philips Color Kinetics ColorBlast 12 floods, 1 x control unit and a multishow unit with event manager/iPlayer3.

The Golden Moon
LEDARTIST (Teddy Lo)

Hong Kong

This page | Appearing aflame with more than 11,000 pixels of RGB LED light, *The Golden Moon* was a spiky, dome-shaped pavilion attracting visitors to Hong Kong's 2012 'Lee Kum Kee Lantern Wonderland' festival at Victoria Park. Measuring 18m (59 ft) high and 21m (69 ft) in diameter, the tensile structure was built above a shallow pool and internally decorated with many floating and aerial Chinese lanterns. Created by LEDARTIST, a pan-Asian lighting agency led by Teddy Lo, the structure was studded with 240 sets of RGB LED luminaires.

Technology | 240 x sets of Traxon Dot XL-6 RGB LEDs, e:cue e:pix video micro converters.

Metro Wraiths

Positive Attracts
Edwin Cheong

Singapore and Sydney, Australia

This page | Glowing like rainbows in human form, nine ghostly 'heroes' stand in military order beside the water at Singapore's Marina Bay. Taller than conventional humans, these wire-framed, fabric-clad, LED-lit lantern figures glow with positive energy that intensifies and vibrates when observers trigger sensors linked to the colour-changing RGB luminaires. Each figure in Edwin Cheong's *Positive Attracts* ensemble is dedicated to a real human hero, identified on a nearby plaque. After first appearing at the 2010 'iLight Marina Bay' festival in Singapore, they were exhibited at the 2011 'Vivid' festival in Sydney.

Technology | 80m (265 ft) Philips Color Kinetics RGB LEDs, IR Motion sensors.

Guardians of Time
Manfred Kielnhofer

Berlin, Germany, and other locations

Opposite page | Shrouded in shining red hoods and floor-length cloaks, mysterious human-akin figures gather conspiratorially in public plazas during evenings of community celebration. These 3m-tall (nearly 10 ft), monk-like *Guardians of Time* convey a sense of immortality and theological authority to protect and enforce universal laws of existence. Created by Austrian artist Manfred Kielnhofer, the guardians have been made in polyester resin, stone and bronze at various sizes and with different fabric cloaks. In recent appearances at European light festivals, some spectators have glimpsed these figures, or just their faces, radiantly glowing (from internally installed LEDs).

Technology | 5 x 3W LEDs, 5 x 12V batteries, 5 x resin figures with fabric cloaks.

Exhibits   **Dynamic Objects**

Orthogonal Twists

A Blue Mirage in the City of Light
WY-TO Architects

Singapore

Opposite page | To contradict the gold-tinged night lighting of Singapore's high-rise buildings, French architects WY-TO (Yann Follain and Pauline Gaudry) designed a light sculpture that radiates stripes of ice blue and cold white (5600K) light. Titled *A Blue Mirage in the City of Light*, their concertina passage also complemented Singapore's water-reflected skyline at the 2010 'iLight Marina Bay' festival. Striped to emulate the city's tall buildings, obliquely arranged acrylic panels included highly reflective outer surfaces and polarized PVB films, diffusing surrounding light sources and generating kinetic, ghostly and mirage-like impressions.

Technology | Philips Color Kinetics eW Flex SLX Flat Lens and iPlayer 3 control system, 24V power, clear acrylic panels sandblasted on three sides, 3M QLF mirroring and polarizing film.

Dar Luz
Ali Heshmati, Lars Meeß-Olsohn

Eindhoven, The Netherlands

This page | Curtains of coloured lasers triggered body scans of people walking precariously along a 22m (72 ft) catwalk through a twisted tunnel clad with Lycra fabric stretched over square wooden portal frames. Detecting the motion and mass of each human, the electronic system caused changes to the colour and intensity of lighting and the pulsing of sounds. Titled *Dar Luz* ('to give light', or 'give birth', in Spanish), this provocative 'social sculpture' was demonstrated by architect Ali Heshmati (LEAD) and light architect Lars Meeß-Olsohn at the 2008 'GLOW' festival in Eindhoven.

Technology | Reactive LED spots, reactive lasers, wooden frame, Lycra Expandex fabric.

198

Botanic Synthesis

Mobile Orchard
Atmos

London, UK

Biomorphic architecture, living structures, digital craftsmanship: these paradoxical visions are becoming tantalizingly feasible to fuse. Prototyping complex visions across these frontiers is London artist-architect Alex Haw (Atmos), who led engineers, designers, fabricators and others to build a CNC-routed, laminated timber, LED-illuminated 'fruit' tree (with edible apples) on the footpath at London's Devonshire Square and two other venues. Designed as an urban landscape-art centrepiece for the 2013 'City of London' festival, Atmos's *Mobile Orchard* 'grew organically', using algorithmic scripts to generate diverse forms that were modelled in precise detail for cutting every slice of wood and translucent white plastic leaves shaped like the area plans of inner London boroughs. Built of spruce, it was designed with removable branches (for transport) and incorporated naturalistic seating around the 'roots' and trunk. Lighting included warm white LED Linear strips set into grooves of plywood, Wibre LEDs used as branch-mounted fireflies and pre-set programming developed with Arup.

Technology | 85m (280 ft) LED Linear VarioLED Venus TV IP67 16mm-wide ($^5/_8$ in) 24V encapsulated warm white LED strips, 50 x Wibre ground-mounted LED fixtures with POWLed, 1 x Traxon S2 Butler DMX controller.

Complexity Constructs

Illumination Disorders II
Tay Swee Siong

Singapore

This page | Voices activate the LED night lighting of a disorderly sculpture assembled from discarded plastic bottles and nylon textiles, creatively cut like colourful flowers and translucent fragments, and held together with metal wires. Singapore sculptor Tay Swee Siong added piezoelectric sensors to some of the flowers, detecting voices to trigger pre-programmed lighting responses via an Arduino microcontroller.

Technology | Arduino UNO sensing electret microphone and control, 500 x RGB LED pixels with WS2812 from Adafruit.

Orkhēstra
M. Hank Haeusler and students

Frankfurt, Germany, and Sydney, Australia

Opposite page | Pioneering 21st-century advances in media architecture, Sydney-based academic and artist Matthias Hank Haeusler led a multi-national group to construct a radiant, coral-inspired, inflatable light artwork for the 2014 'Luminale' festival in Frankfurt (then shown at Sydney's Powerhouse Museum). Its organic form and dynamic lighting sequences arise from experiments with 3D topological, mesh geometries and bespoke software controlling the interactivity of each LED.

Technology | 7,732 x uniquely laser-cut perforated pieces of 1.2mm-thick ($^1/_{32}$ in) polypropylene, 5,500 x AHL LED S25-3 pixels.

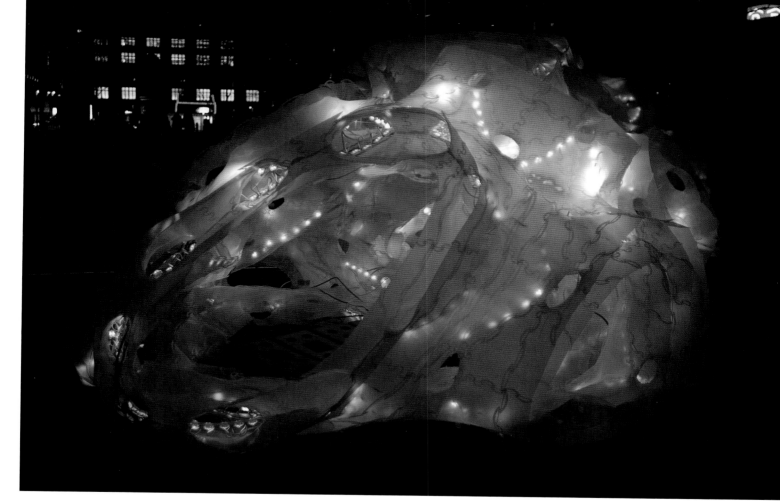

Exhibits

Dynamic Objects

Cycloid Sighs

Lotus Dome
Studio Roosegaarde

Lille, France, and other locations

Updating Buckminster Fuller's futuristic geodomes, yet reminiscent of ancient Moorish filigree lanterns, the *Lotus Dome* by Dutch artist Daan Roosegaarde is a 'living organism' comprising hundreds of Mylar foil petals which gently curl open or closed in sensor-triggered responses to human proximity and touch. Internally lit with LED lamps (creating an intricate pattern of triangles around the edges of the petals), the dome measures 3m (10 ft) in diameter and 2m (6½ ft) high. The concept integrates themes from nature, architecture, light art and electronic (remote sensing) technologies. Originally exhibited in 2011 inside the Sainte Marie Madeleine Church in Lille, France, it has since been shown in other cities in Europe, Israel, the United States, China and the Middle East.

Technology | LEDs, sensors, control system, Mylar sheets.

Delicate Subtleties

C/C
Angela Chong

Singapore, Sydney, Australia, and Amsterdam,
The Netherlands

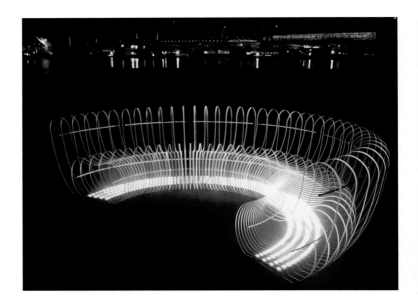

This page top | Intriguingly immaterial yet comfortably familiar, the C/C public outdoor seat, by Singapore artist Angela Chong, glows with a ghostly radiance. Made of light-refracting clear acrylic sheets bolted with stainless steel screws and spacers, it is uplit by Philips Color Kinetics iColor Flex LMX LEDs fitted around its C-shaped base. Street lighting for India's traditional Diwali light festival inspired part of the colour programming, using Philips iPlayer3. This work was originally created for the 2010 'iLight Marina Bay' festival in Singapore, then was exhibited at 'Vivid' Sydney and the Amsterdam Light Festival.

Technology | 350 x nodes of Philips Color Kinetics iColor Flex LMX, Philips iPlayer3 colour programming, 79 x acrylic sheets.

Flower from the Universe
Titia Ex

Various locations

This page bottom and opposite | The long, elegant petals on a tropical passiflora bloom inspired Amsterdam light artist Titia Ex to develop her sophisticated *Flower from the Universe* light- and colour-responsive, sensor-activated artwork. Shown at festivals across Europe since 2010, this digitally enabled organism comprises a central 'heart' with eighteen branches emulating synapses of a human brain. This is surrounded by a garland of 35 LED-illuminated stems, radiating outwards in two layers, like petals. The synapses and the stems are illuminated with clear polycarbonate LED strips that can change colour via a pre-programmed pattern, or in response to nearby colours and movements detected by sensors in seven pods. This capability allows the light flower to 'interact' with its local environmental conditions. It can be operated via a time-switch and is elevated by a stainless steel sub-frame with four adjustable legs.

Technology | Osram flexible RGB LEDs, Philips RGB LED strips, clear polycarbonate, sensors.

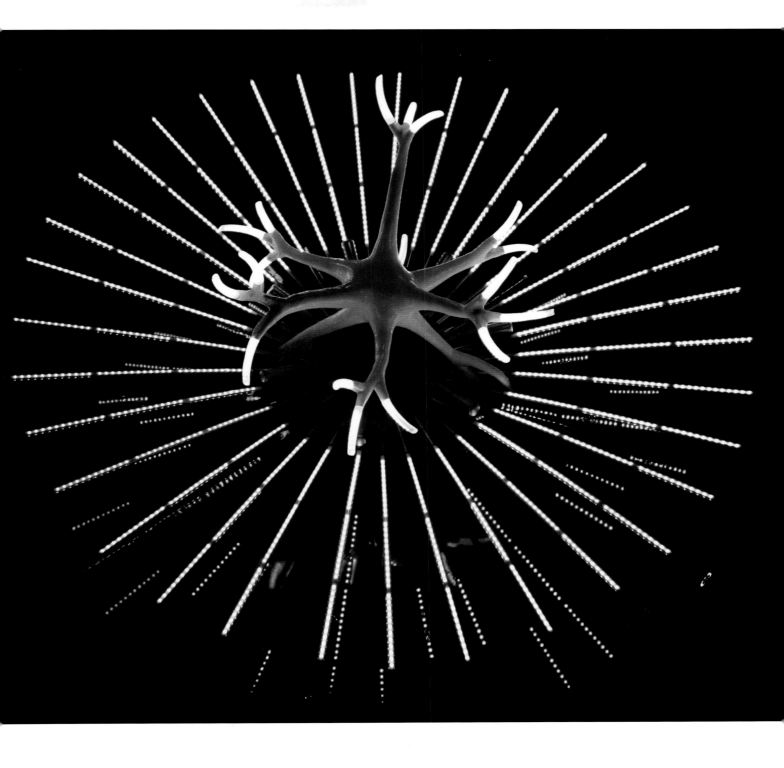

Dynamic Objects

Symbolic Gestures

e|MERGence
The Buchan Group

Sydney, Australia

This page | Lying askew, seemingly discarded in Sydney's busy Martin Place, a giant, multi-faceted head comes to life for spectators at night. Architects and designers at The Buchan Group used advanced systems of computer modelling, digital fabrication, 3D scanning and video projection mapping to upset conventional ideas about the look and positioning of faces in public life. During the 2014 'Vivid' festival, cameras recorded viewers watching their *e|MERGence* interactive performances, then 'topographically' mapped these video captures onto the facial facets of the sculpture.

Technology | 4 x Optoma 5000 lumens HD projectors, Microsoft Kinect for PC (motion capture), Cisco and Logitech HD webcams (video capture), Røde and Sennheiser microphones (audio capture), media server with TouchDesigner user interface, Autodesk 3ds Max, Adobe After Effects and Element 3D.

Light of the Merlion
Ocubo (Nuno Maya, Carole Purnelle)

Singapore

Opposite page | Singapore's statuesque national fountain, the Merlion (head of a lion, body like a fish), was transformed from serene white marble to bearer of dazzling night costumes of multi-coloured light. At the 'iLight Marina Bay' festival in 2012, Portuguese artists Ocubo set up an interactive 3D projection mapping that allowed the public to change the Merlion's patchwork light colours instantly via its image on a nearby touchscreen.

Technology | 2 x Panasonic 15K lumens projectors, 1 x interactive touchscreen laptop, 1 x interactive multimedia application, video mapping software.

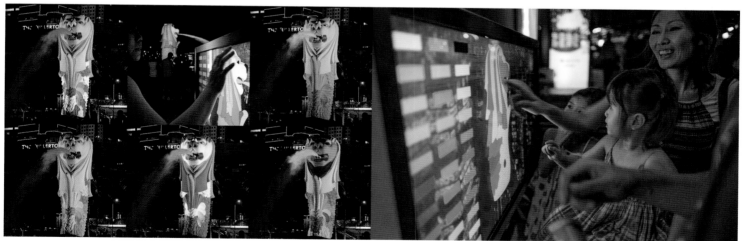

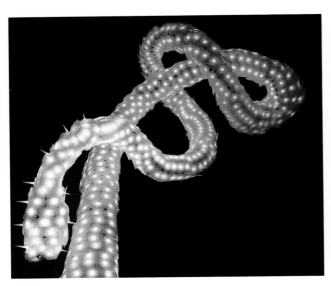

Fluid Creatures

Luminous Organisms

LEDARTIST (Teddy Lo) / Caitlind R.C. Brown and Wayne Garrett / Pascal Petitjean, Aamer Taher and Simon Lee / Jason Peters / Sun-Yu Li

Various locations

Opposite top left | *Techno Nature: Bacillus*, by Teddy Lo, represents a giant single-cell organism that expresses 'moods' (including fear, joy and fuss) when humans approach.

Technology | RGB LED strings, LEDs, silicon pixel cover, sensor, computer control system, DMX decoder, LED driver, metal frame.

Opposite top right | *Cloud* is a globally nomadic, interactive, cumulus-styled canopy of expired incandescent bulbs that diffract the glow of 200 active LED bulbs. Created by Caitlind R.C. Brown and Wayne Garrett, it includes silver chains falling like rain. These can switch on lightning effects.

Technology | 6,000 x incandescent light bulbs (new and burnt out), 200 x LED bulbs, electronics, pull chains, steel, adhesive.

Opposite bottom | *Jellight* is a cluster of 'jellyfish' conceptually emerging from polluted water to float to the heavens. Made with internally lit helium balloons designed by Singapore architect Aamer Taher and made by Pascal Petitjean of Airstar with Australian film light expert Simon Lee, these 6m-tall (20 ft) creatures include 'tentacles' of luminous white LEDs.

Technology | Airstar helium balloons and 'tentacles' made with Teflon fabric, LEDs installed internally.

This page top | *Andrea*, by Jason Peters, is one of a series of sublime, phosphorescent eels assembled from polyethylene buckets, internally illuminated with LED ropes. These night-radiant creatures undulate through trees, or scaffolds of fine steel wire.

Technology | 300 x 3.5GL white buckets, screws, LED rope lights.

This page bottom | *The Beginning*, by Sun Yu-Li, was inspired by DNA spirals and a boast the artist attributes to science philosopher René Descartes: 'Give me a dot and I will revolve the world.' Made with stainless steel and LED tubes, it was a festival exhibit under a banyan tree outside the National Museum of Singapore.

Technology | Stainless steel tube twisted into shape and cut with dense slots on the surface, 6 x linear LEDs inserted inside tube.

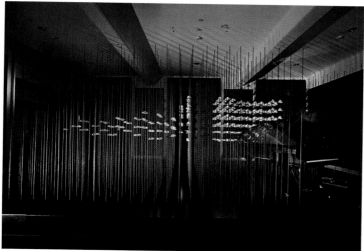

Memento Mori

Urban Light
Chris Burden

Los Angeles, USA

Opposite top | More than 200 vintage streetlamps are closely arrayed in long colonnades on the Wilshire Boulevard forecourt of the Los Angeles County Museum of Art (LACMA). Artist Chris Burden created *Urban Light* as a memento mori of industrial-era road infrastructure. From 2000 to 2008, he collected several hundred discarded 1920s–1930s California streetlamps in seventeen styles, most including fluted poles. These were sandblasted, sprayed with grey paint and converted to electricity, then installed as a cohesive homage to one genre of relics of last century's civilization.

Technology | 202 x cast-iron streetlights, restored and electrified.

Daylight Flotsam Venice
Bill Culbert

Venice, Italy

Opposite bottom | New Zealand-born, France-based Bill Culbert exposed the increasing obsolescence of ubiquitous 20th-century manufactured goods with his 'Front Door Out Back' exhibition, produced for the New Zealand pavilion at the Venice Biennale of 2013. After decades of experimentation with the phenomenological and optical characteristics of electric light, he scattered fluorescent tubes and plastic bottles across a canal-flooded brick chamber of the Santa Maria della Pietà palazzo, discarding these products as debris drifting into history.

Technology | Fluorescent tubes, plastic vessels, fittings.

Doves That Cry
Mary-Anne Kyriakou, Joe Snell

Sydney, Australia, and Singapore

This page | A flock of radiantly white aerial objects – representing doves, the universal symbol of peace – hover in dark space above a polished black piano. On inspection, the 'doves' are reflectors fitted with white and blue LED lamps, causing eerie reflectance effects. Sometimes a pianist plays a melancholy melody, highlighting fears of mass extinction of species.

Technology | 150 x recycled reflectors from Euroluce, 150 x Philips Color Kinetics LEDs, Philips Dynalite control system, custom electronics and sound sensors.

Immersions, Interactions

Everyone alive is immersed within – and interactive with – light from our planet's sun. But with semiconductor-enabled LEDs and sensors, there are almost infinite possibilities for designing artistically exciting user interactions and experiences (UI and UX).

LED lamps are common outputs and nodes for electronics experts building circuits and meshes of synchronized equipment for public multimedia displays. Common user-triggered lighting thrills include LED spectacles responding to bass players at rock concerts and people jumping and dancing in front of cameras and hidden sensors to see themselves perform on giant screens.

Ocubo, a Portuguese light art team, create urban touchscreen games whereby children can operate a computer screen to 'paint' a city monument with 3D-mapped light in RGB colours. Examples include creating coloured patchwork projections on Singapore's riverside Merlion statue (see p.206) and remotely playing pinball with graphic light projections onto Jerusalem's historic Damascus Gate (see p.214).

Human-computer interfaces (HCI) underpin a new digital urban creative movement known as 'social light', through which people enjoy or learn from digital incidents and events that can be preset, spontaneous or both. These can happen in physical or virtual environments, or a combination of the two.

A key trend is (tele)presence – whereby users feel they are 'really there' in a virtual environment, forgetting they are wearing mediation devices such as Oculus Rift headsets or sensor gloves. Presence is the most extreme form of immersive art.

UI and UX episodes are always enabled by coded manipulations of data streaming online between diverse devices. This general realm of technology has been labelled 'the internet of things' (IoT). Controversial aspects of the movement include the ability for government agencies – or hackers – to monitor precisely and remotely control uses of electric appliances connected to any public power supply.

IoT specialists – including many light and sound artists – like to invent novel demonstrations with remote sensing and electromagnetic wave-imaging technologies such as RFID (radio-frequency identification), outdoor laser scanners, biometrics, magnetic strips, projectors, drones, and 'audio and visual' signals.

Where are the main fountains of interactivity innovation? Research clusters at the Massachusetts Institute of Technology – notably MIT's Media Lab, SENSEable City Lab and former Smart Cities Lab – have been leading the world in producing valuable and globally replicable IoT applications. Many European researchers belong to the Austria-based International Society for Presence Research. A team at Italy's former Interaction Design Institute Ivrea invented the popular open-source Arduino processor board, which is now routinely used for temporary urban light interactions. Also significant is the annual Prix Ars Electronica competition, which attracts dozens of radical proofs of concept to the awards ceremony in Linz each September. In addition, crucial hardware products, with sophisticated operating systems, regularly come from the in-house research teams at major equipment producers in Asia and Europe and on the US West Coast.

Global lighting corporations such as Philips, TRILUX, Osram, Panasonic and Zumtobel have formed the Zhaga consortium to share specifications to maximize cross-uses of their mass-produced LED lamps. All lighting suppliers are improving their 'kit' (equipment) to enable automatic responses to pedestrian movements, moonlight levels, wind or rain conditions, and specific hand gestures.

Many of these non-lighting advances contribute to the smart light movement, which promotes both renewable sources of energy and energy efficiency using automatic digital controls (rather than human memory to flick on-off switches manually).

Also relevant is the maker trend to craft electronic objects and gadgets using both sophisticated electronics and found parts. Indie artists are glueing, screwing and gaffer-taping LED bulbs and tubes into networks of cables, internet protocols and components such as Arduino boards and error-sensing servomechanisms (servos). These unstyled tools can be controlled with touchscreens or keyboards on one or more mobile devices, often using online apps – software programs with custom-designed graphical user interfaces.

One of the world's most sophisticated practitioners developing interactive light and data art works is Montreal- and Madrid-based Mexican artist Rafael Lozano-Hemmer. He mashes robotics, cell phone interfaces, video and ultrasonic sensors, projections, positional sound, and other tools and techniques to offer 'platforms for public participation' in parks and public galleries of major cities. In his 'Monument/Antimonument' speech at the 2014 Sculpture City conference in St Louis, he said:

The work is a platform for public self-representation. Yes, the platform has an authorship, but it depends on participation to exist: it has points of entry, loose ends, tangents, empty spaces and eccentricities. All art has awareness: pieces listen to us, they see us, they sense our presence and wait for us to inspire them, and not the other way around... It's not what they are, but what they are becoming.

Like many contemporary interactivity experts, Lozano-Hemmer began his career with a tense choice of professional directions between science and art. Inventors who excel across both domains have been rare and often extremely valuable to society's evolution. Their concepts and triumphs tend to stimulate tangents and advances from later generations of talents.

Both interactive and immersive artworks require cross-flows of energy between the participants and key interfaces of the work. A critical difference is that, while interaction demands active decisions and physical movements, immersive light art can require players to assume a mode of stasis ... allowing the potential of what Korean art scholar Jinsil Seo has called 'a transcendent state of consciousness, the Kantian sublime, or a hypnagogic state of reverie'.

Façade Frolics

Wall Pinball: Save the Ocean
Ocubo (Nuno Maya, Carole Purnelle)

Sintra, Portugal

This page top | For Sintra's 2011 'Lumina' festival, Ocubo designed a pinball game to be controlled with a seemingly conventional pinball table, but with a virtual ball projected across the building wall and bouncing off its real obstacles, such as window frames. The game was named after a free online interactive app created by Ocubo to accompany their book *The Plastics*, which presented stories and characters made with drifted plastics salvaged from beaches around the world.

Wall Pinball: The Weather
Ocubo (Nuno Maya, Carole Purnelle)

Jerusalem, Israel

This page bottom and opposite | Played across the Holy City's ancient Damascus Gate during the 2012 'Light in Jerusalem' festival, Ocubo's *The Weather* interactive pinball game required players to control the ball as usual via levers on the game table, but the effects of their actions could be seen vertically and enlarged on the city's walls. Each player was provided with six virtual balls, which 'bounced off' real architectural obstacles.

Technology | Both projects: 2 x 15K lumens video projectors, 1 x interactive pinball table mock-up, customized 3D mapping software, interactive application created by the authors.

Immersions | Interactions

216

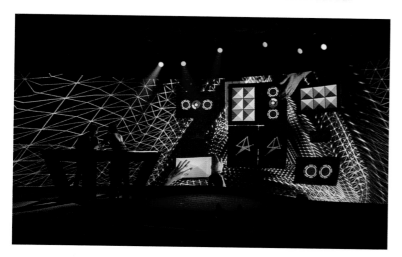

Calls to Action

Biorama
Moment Factory

Montreal, Canada

Opposite page | Orientation is an important concept in designing public museums and galleries. However, the strategy of disorientation was adopted by Canadian multimedia studio Moment Factory for its design of the introductory zone at the Biodome of Montreal. Visitors enter a human-scaled kaleidoscope of animated, mirror-reflected wall and floor projections of brilliantly coloured flora and fauna. Images by the Biodome's photographer, Claude Lafond, are enhanced with synchronized music composed by Vincent 'Freeworm' Letellier. The architectural concept is a triangular room, spatially intersected by a double-sided mirror wall.

Technology | 8,500 lumens video projector, printed RGB vinyl floor, printed RGB vinyl flooring, 12 x PAR RGB LED lamps, 8 x speakers, 2 x sub-woofers, QLab control system, 10 x mirror panels, laminated PVC projection screen, two-way laminated glass panel.

Qualcomm's Uplinq Launch
Moment Factory

San Diego, USA

This page | How to memorably kickstart a semiconductor manufacturer's new product presentation for mobile app developers? California telephony components manufacturer Qualcomm called Moment Factory to create a dazzling opening show for its Uplinq event in 2013. Two electronic music performers – Vincent Letellier (Freeworm) and Jennifer Daoust (Video Girl) – used tablets and smartphones to mix music and video projected live on a giant screen. The soundscape for this enhanced VJ/DJ set was composed by Vincent Letellier.

Technology | 8 x Sony Xperia Z tablets, 2 x LG Nexus 4 mobile phones, 10 x Christie 18K lumens projectors, 1 x 70" flat screen TV, custom VJ booth finished with 2.5m (8 ft) LED strip, 1 x camera.

Playing Around

Panorama
Andreas Groth Clausen, Jakob Kvist

Copenhagen, Denmark

To open Copenhagen's 2013 'Strøm' electronic music festival, its producers created *Panorama* as a memorable, immersive, multimedia artscape in the city. Produced by Andreas Groth Clausen with lighting designer Jakob Kvist and UK-based 3D soundscape artists Illustrious, the *Panorama* concert integrated live electronic and classical music and television-like spectacles of shadowy musicians in windows pulsing with coloured LED lighting. The donut-shaped architecture of the seven-storey Tietgen student housing complex, designed by Lundgaard & Tranberg, allowed the entire performance to be experienced in the round – but in a contra-Shakespearian mode, whereby the audience was located 'on stage', surrounded by the performers.

Technology | From LITECOM: 144 x Martin Stagebars, 50 x Showtec Spectral CYC 650, 35 x Etan 24x3W RGB full-colour LED lights, 60 x Martin moving head lights comprising 16 x MAC II Profile, 8 x MAC Viper Profile and 36 x XB Wash luminaires, grandMA full-size DMX control system.

Immersions | Interactions

Feedback Tactics

Dune
Studio Roosegaarde

Rotterdam, The Netherlands, and other locations

Opposite page | Emulating natural paths lined with flowering bushes, Dutch designer Daan Roosegaarde's *Dune* walkways are electronic nightscapes of human-responsive digital sounds and lighting. Exhibited widely around the world since 2007, and permanently installed along the Maas River, Rotterdam, they include dense clumps of white LED-tipped optical fibres, which are sensor-activated to brighten and fade as wanderers approach and move on.

Technology | Hundreds of fibres, LEDS, sensors, speakers, interactive software and electronics installed as a modular system variable in length.

Desire of Codes
Seiko Mikami

Tokyo, Japan

This page | Visitors wandering through Seiko Mikami's *Desire of Codes* exhibitions are constantly tracked by ninety wall-mounted surveillance cameras and six ceiling-mounted robotic search arms. The video recordings are digitally converted into one circular image, comprising 61 hexagonal facets, like the complex dynamics of a mosquito's eye. Each cell shows a different sequence of visitors' movements captured by the cameras.

Technology | 90 x wall-mounted robotic metal 'tentacles' with surveillance cameras, 6 x ceiling-mounted robotic search arms with video cameras and laser projectors, networked computer and bespoke software, circular projection screen with 61 hexagonal facets.

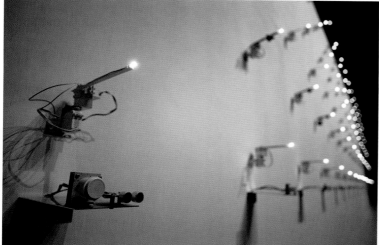

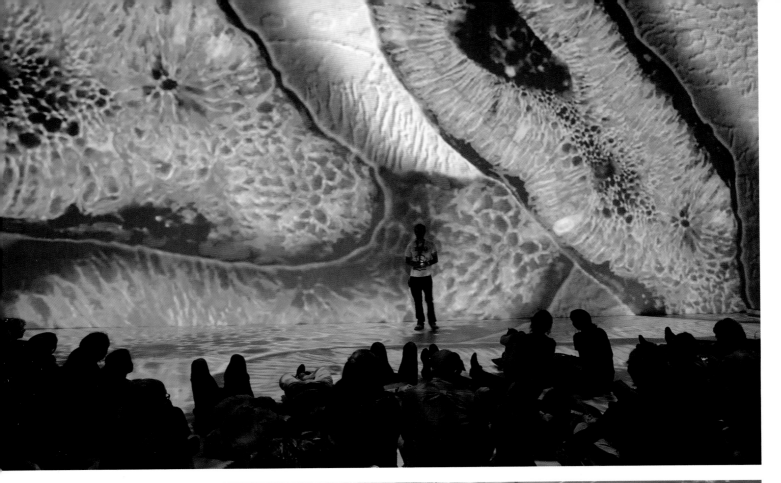

Liminal Zones

Deep Space
Ars Electronica Center

Linz, Austria

Opposite top | Extraordinary immersive experiences are granted to small audiences within the 'Deep Space' chamber of the Ars Electronica Center in Linz. The studio incorporates a 16 x 9m (53 x 29 ft) infinity cyclorama with high white wall and floor viewing screen, a 5m-high (16 ft) platform for overviewing certain floor performances, and eight active stereo-compatible projectors that can be coordinated for seamless and precise 3D imagery.

Technology | 8 x Barco Galaxy NH12 HD projectors, 5.2 sound system, bespoke Ars Electronica control system.

Voyage of Discovery
Ars Electronica Futurelab

Wattens, Austria

Opposite bottom | New staff joining crystals supplier Swarovski are offered remarkable induction sessions in a virtual reality chamber at the company headquarters in Wattens. The experience was designed by electronics engineers at Austria's Ars Electronica Futurelab with digital pens and responsively coded wall and table surfaces by Anoto of Sweden.

Technology | Anoto digital pen and responsive pattern-printed book as control device, 16 x projectors for seamless panoramas on walls and an interactive table, content management system.

Amazonas: A Media Opera: Part III
ZKM

Karlsruhe, Germany, and other locations

This page | Multimedia artists at Germany's ZKM institutes created the third (final) part of an international 'media opera' highlighting Brazil's ultra-sensitive Amazon region as a microcosm in the geopolitics of today's global climate crisis. For *Amazonas Part III: In anticipation of a rational method to solve the climate crisis*, ZKM's directors projected imagery based on computer calculations across an assembly of small cubes, simulated a climate change conference, and digitally involved the audience in a 'Greek' chorus forecasting human extinction.

Technology | 2 x Apple Mac Pro computers, Yamaha M7CL mixing console, 28 x Meyersound speakers, Zirkonium Sound Dome software, Ableton ONE sequencer, time code synchronization with video computers.

Vocal Projections

Megaphone
Moment Factory

Montreal, Canada

Opposite page | Montreal's French- and English-speaking citizens rebooted Britain's spontaneous commentary tradition of 'Speakers' Corner' during several autumn nights along the Promenade des Artistes at the Quartier des Spectacles. Designed by multimedia studio Moment Factory and sponsored by the Quartier des Spectacles with the National Film Board of Canada, the event involved participants speaking into a microphone, their statements being amplified to the audience by a pole-mounted megaphone, and bespoke voice-recognition software transforming the words into images for projection onto the President Kennedy building. The 'Speakers' Corner' was styled like a music concert stage, with a three-storey stack of graphically painted shipping containers, while a 'Speakers' Walk' highlighted seven of Montreal's significant public commentators.

Technology | 9 x Christie 18K projectors, VYV Photon playback system, Shure 577B microphone, 4 x VL3000 mobile lights, CRIM bilingual speech recognition software, Derivative TouchDesigner generative content creation, Max MSP audio analysis software, megaphone, shipping containers.

Voice Tunnel
Rafael Lozano-Hemmer

New York, USA

This page | To celebrate its 2013 Summer Streets promotion, New York's Department of Transportation commissioned Rafael Lozano-Hemmer to develop a voice-activated public art experience along the Park Avenue Tunnel. He dramatically uplit the 427m-long (1,400 ft) arched interior with 300 high-powered theatre spotlights, then allowed visitors to flash these arches of light with their own phrases and rhythms of speech. Participants spoke, one at a time, into an intercom placed in the middle of the tunnel. Each voice generated a Morse-like code of electric pulses linked to a specific group of floor-mounted spotlights and synched loudspeakers. In maximum mode, 75 different voices caused different sounds and light-blinking patterns along the tunnel – to be sequentially replaced by new voices generating different atmospheric effects.

Technology | 300 x 750W Source Four spotlights, ETC dimmer racks, 150 x loudspeakers, Socapex power cable, fibre optic cable, XLR cable, 4 x 500 Amp generators, microphone, custom hardware and software.

Motion Effects

Giant Bicycle Headquarters
CMA Lighting Design

Taichung, Taiwan

This page | Dynamic fringes of radiant dots border the ceilings of balconies curving around the headquarters of Taiwan bicycle maker, Giant. Designed by Taipei-based CMA Lighting, this scheme sandwiches a white upper layer printed with black dots, a lower perforated metal layer, and blue LED linear lights (matching Giant's brand colour) in the 20cm (7¾ in) cavity in between. The black and blue dots pulse in response to people walking underneath. Other lighting tactics include motion sensors triggering step lights when cyclists ride up the entry ramp, and white LEDs for entry downlighting and footpath poles.

Technology | 340m (1,115 ft) blue linear LEDs (27W/m, 38°), 27 x 2.5W step lights with motion sensors, 18 x recessed multi-lamp downlights (5W and 7W PAR30), 4 x custom streetpoles, each with two 4200K 70W ERCO HIT lights (one 7° spot and one 85° flood), from GL Lighting.

CYCLE!
CLOUSTON Associates

Sydney, Australia

Opposite top | Five large aluminium rings, fitted with blue LED strips, encircle the trunks of two trees. Visitors pedal nearby exercycles to activate the circles of light. This work was strategically sited on the foreshore, with overlapping violet curves surging in reference to historic tidal rhythms. It also highlighted the importance of physical exercise.

Technology | 15-20m (50-65 ft) 12V Neo-Neon blue LED tubing, recycled washing machine induction motors, 5 x modified BMX bicycles on stands, waterproof transformers.

V.I.P.
Aleksandra Stratimirović

Uppsala, Sweden

Opposite bottom | At the Heidenstam school in Uppsala, evening visitors can pose on a circular 'arena' of gloss white paint, illuminated by overlapping light beams of red, blue and green, projected from three spotlights on equi-distant poles.

Technology | 3 x WE-EF FLC240 Zoom Spot spotlights, white paint, Tehomet Conical 8m-high (26 ft) steel poles.

Flame Swirls

Fractal Flowers
Miguel Chevalier

Céret, France, and other locations

French digital artist Miguel Chevalier creates 'virtual gardens', 'artificial paradises' and 'vegetal universes' of gigantic, fantastic, stylized blooms. Although obviously artificial in design, they seem to grow, fade and move realistically on screen walls in response to sensor-detected movements of viewers. Chevalier is one of several leaders of today's advances in algorithmically generated, evolutionary computer art, visualizing nature's biological processes through the mathematical principles of fractals and other scientific theories of metamorphosis. This 2014 showing of his *Fractal Flowers* was part of his 'Artificial Paradises' exhibition at the Museum of Modern Art in Céret, France, following earlier exhibitions in Hong Kong, Lisbon, Brasilia and Seoul.

Technology | 2 x Epson 5000 lumens video projectors, 1 x infrared sensor, custom software on Linux.

Interior Intrigues

What is public space? Cultural and political scholars are teasing the social connotations of that Marxist question, as boundaries blur between private, commercial and government land ownership.

When we sip coffee under a new glass canopy on the roof deck of a historic stone post office, do we know (or care) who owns the particular square metre of ground (in legally transferred airspace) that our feet right now are touching? If we steal someone's handbag in a shopping mall, does it matter that a private agency pays the salary of the security guard who restrains our liberty before the police arrive? These are touchy matters in academic debates.

City governments increasingly allocate long-term BOOT (build-own-operate-transfer) contracts to entice commercial developers into once-taboo public-private partnerships (PPPs) to transform derelict docklands, industrial zones and heritage buildings for 'adaptive reuse'.

Astute leaders of major developments now see light art as a priority strategy to activate new potentials for old land assets. As well as installing prominent outdoor fixtures, they favour light art to help brand and make memorable the indoor courts and skylit atria of shopping malls, hotels, office buildings, and public museums and galleries.

Obviously skylights are installed to flood interiors with daylight and are not effective at night, when creative installations using artificial light are intended to shine. But how many people will be walking through a public interior after business hours? Office foyers and government buildings are usually locked after 6pm, and then these zones are not 'public' interiors.

However, spectacular light sculptures, perhaps suspended from high foyer ceilings or installed on walls around central lift cores, can be remarkably effective after-hours signals of building prestige. Although confined indoors, these can be viewed through glazed walls by peak-hour commuters. As night descends, they offer sparks of splendour to citizens in transit – especially dazzling on gloomy or rainy evenings.

Most shopping malls and airports are artificially lit to pseudo-daylight levels at all hours, but they have high foot traffic and are logical domains for global corporations to promote their products and services via diffused-light poster boards and high-res video displays. These may show simple advertisements or conceptual imagery by the world's most inventive media artists.

Theatres, concert halls and casinos have always been generous patrons for arresting art light effects designed to be viewed indoors. For foyers, the operators favour backlighting of 'monumental' walls clad with panels of fine polished marble, fairyland lighting of internal trees, large pixel screens or telescreen grids on prominent walls, large illuminated sculptures and water features, and sculptural wall or ceiling murals with LED lamps or tubes providing highlights.

Nightclubs (whether regular commercial operations or warehouses hosting one-night special events) are dark containers for crowds to share extreme immersive experiences. Electricity-facilitated streetlife sparked the early jazz clubs of Paris, London, Vienna and Berlin during the 1920s – and, since the arrival of electric guitars in the 1960s, music venues have been playgrounds for talented sound and light artists to thrill young audiences to heights of nervous energy and sensuality.

Sophisticated immersive sound and light experiences are created by teams of electronic artists – for example, Stuttgart-based 'Visual Piano' exponent Kurt Laurenz Theinert and New York-based Ursula Scherrer – who travel globally to perform 'intimate landscapes of fleeting sensations' for city cultural celebrations and festivals.

Less immersive, more minimal, but deeply provocative are conceptual art rooms, such as Seiko Mikami's *Desire of Codes* (see p.221), in which visitors are nonplussed to realize that their meanderings are instantly followed by arrays of whirring security cameras. In a closed, dark room, a ubiquitous condition of the city becomes hyperreal and impossible to ignore.

Another notable 'multisensory' installation is *Between | You | And | Me* by Anke Eckardt, where viewers in a confined space are disturbed by sounds of crashing glass when they unwittingly 'break' an invisible 'wall' of ultrasonic and light beams.

Astonishing, too, are intimate exposures to the complex and feather-delicate nocturnal ecosystems created by Canadian architect Philip Beesley (see p.244). Not fully immersive or interactive, but slightly sensor-responsive to movements by visitors, these are suspended compositions of fine glass tubes, LEDs, sensors, polymers, wires and meshes. High-tech interpretations of dense, primordial plantscapes, they are today's most magical demonstrations of the oxymoronic concept of 'metabolic architecture', fusing contrasting qualities from nature, architecture and electronics.

Ceilings in public spaces offer exciting potentials for designers of luminous art – if their teams can resolve the technical challenges. Before Edison's incandescent lamps, chandeliers were the ultimate luminous spectacles for palatial interiors, refracting wax candles into light beams like a thousand jewels. Because LED lamps are small, cool when lit, safe to touch, long-lasting, energy-efficient and inexpensive to operate, they are a godsend for retro-fitting historic monuments. They enable regular contemporary use of even the most fragile chandeliers, and are enabling a magnificent renaissance in creating interior scenes of glamour, mystery and wonder.

Atmospheric Explosions

Ylem
Atelier H. Audibert

Lyon, France

In the late 1940s, American scientists invented a theory that primordial plasma, named 'ylem', was transformed into subatomic particles and elements during the Big Bang. This concept influenced the explosive *Ylem* light shows performed by French artist Hervé Audibert during the December 2013 'Fête des lumières' in Lyon. The *Ylem* shows, involving 150 rotating and colour-changing light beacons, dramatically 'activated' the glassy architectural shards of the new Musée des Confluences (sited at the Rhône and Saône river junction), designed by Austrian studio Co-op Himmelb(l)au with façade engineers VS-A Group.

Technology | 150 x rotating and colour-changing light beacons, control system.

Le Département du Rhône construit le musée des Confluences. Ouverture fin 2014. RHÔNE
LE DÉPARTEMENT

Light Showers

Rain Room
Random International

London, UK, and New York, USA

Trusting invisible technologies to keep dry, gallery visitors wander calmly through 100 sq m (1,075 ft²) of dense, cacophonous rain. Participants are sheltered from the drops falling directly above them by a system choreographed using sensors, camera tracking, custom software and water management. The 'rain' is continuously recycled through a grated floor and injection-moulded ceiling tiles. Introduced at London's Barbican (The Curve) in 2012, then exhibited at New York's Museum of Modern Art in 2013, this challenging artwork dramatically manipulates perceptions of weather and atmospheric conditions inside a contemporary art gallery. Random International's founders – Stuart Wood, Florian Ortkrass and Hannes Koch – are conceptual artists, scientists and engineers based in London and Berlin. They use computational systems to express dynamic interplays of light and matter, often contrasting reflections, transparency and solidity in monochromatic works.

Technology | Custom software, 3D camera tracking, water management system, solenoid valves, pressure regulators, water pump, computer.

Virtual Floors

Magic Carpets
Miguel Chevalier

Andria, Italy

Opposite page | Centuries before digital images composed of pixels, artists around the Mediterranean created intricately elaborate decorations using mosaic tiles, tapestry knots or stitches. French artist Miguel Chevalier reinterpreted those ancient art precedents with his complex designs for virtual carpets, projected across the octagonal courtyard of southern Italy's 13th-century Castel del Monte. Now preserved as a UNESCO World Heritage site, the castle's architecture includes sophisticated geometric calculations aligned to medieval theology and astrology. Chevalier's 'magic carpets' also incorporate pyschedelic and other cinematic effects alluding to biomorphology and utopian dreams of escape from reality. These auto-generative, interactive visual sequences were partially distorted by the footsteps of observers and synchronized to music by Jacopo Baboni Schilingi.

Technology | 2 x Panasonic 12K lumens video projectors, 1 x infrared sensor, 4 x speakers, custom software on Linux, 1 x computer.

Tapis Magiques (Magic Carpets)
Miguel Chevalier

Casablanca, Morocco

This page | Sinuous and sensuous swirls of bright colour undulate along the nave of the ancient Sacred Heart church in Casablanca, Morocco. Reacting to footsteps and synched to music by Michel Redolfi, this giant carpet of computerized projected light was performed by Miguel Chevalier for a Heritage Days cultural promotion in 2014. Reminiscent of Arabian Nights fantasy stories, the digital imagery constantly evolved in biomorphic patterns auto-generated by software using cellular automata principles.

Technology | 3 x Panasonic 20K lumens video projectors, 3 x infrared sensors, 8 x speakers, custom software on Linux, 1 x computer.

Transit Systems

NOVA Display System
Martina Eberle and Christoph Niederberger

Zurich, Switzerland

This page | Suspended above teeming crowds at Zurich's train station, NOVA was the world's first LED voxel 3D digital screen display system. Invented by Martina Eberle and Christoph Niederberger, with collaborators on hardware and software development, the system comprised LED modules with 1m-long (3¼ ft) voxel strings, controlled by bespoke software. To compensate for the low resolution of the physical system, NOVA sculptures display content in real three dimensions, place objects in a physical 3D space defined by voxels, and display videos projecting 2D content into the 3D space. Content can be playlisted or created in real time, using various applications through interfaces such as touchscreens, mobile phones and microphones.

Technology | 2,500 x strings of colour-changing RGB LED voxels, bespoke software.

Tom Bradley International Terminal
Moment Factory

Los Angeles, USA

Opposite page | Reigniting the romance and adventure of travel to foreign lands, seven kinds of immersive, interactive, video-rich structures stimulate travellers walking through the new Tom Bradley Terminal at Los Angeles Airport. These walls, portals and towers were produced by MRA International and Sardi Design, with content produced by Moment Factory. The tallest structures are *Time Tower*, a 22m-high (72 ft) screen-clad clock tower around the main elevators, with a touch-reactive surface affecting video content across 600 sq m (6,460 ft²) of LED surfaces, and two *Concourse Portals*, each comprising ten screen-clad vertical panels, all playing audio-visual content responding to data on nearby pedestrian and flight traffic. The four wall works are the *Welcome Wall* (24m-high; 79 ft), the *Bon Voyage Wall* (slow-mo filmic images inspired by Philippe Halsman's 'Jumpology' series), the *Story Board* (36.6m-high; 120 ft), and the *Destination Board* (a generative video 'data cloud' display of flight information and destination imagery).

Technology | 8 x X-Agora servers, 20 x X-Agora media players, 4 x X-Agora interactive managers, 1.9m sq m (20½ ft²) LED surfaces (1.9 million pixels), 30 x multimedia loops of 3-9 mins each.

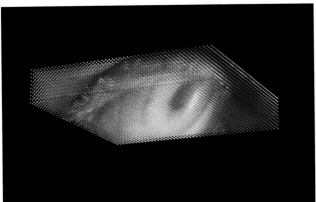

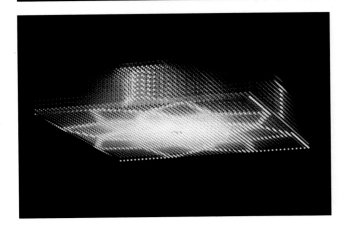

Exhibits   Interior Intrigues

Commercial Stimulus

Mirrors
Random International

London, UK

This page | Behaving like a swarm of living organisms, 180 mirrors twist on multiple slants to catch, transmit and distort reflections around the atrium of London's Berkeley Square House. Each mirror can move autonomously, but they also collectively respond to sensors tracking people in proximity.

Technology | 180 x mirrors, motors, servos, motion-tracking software, networked control system.

Oakley Ceiling Screens
Moment Factory

New York, USA

Opposite top | For Oakley's flagship store in Manhattan, Moment Factory developed a multi-screen ceiling with video loops promoting Oakley-sponsored athletes and brand imagery. The Samsung LCD screens are arrayed in nine banks of three within a frame of aluminium-faced panels, folded and scored with CNC machines. These are optically arranged to provide a dramatic composite image when seen at the entrance, then to seem more fragmented as viewers walk to the back of the store.

Technology | 27 x Samsung professional TFT LCD screens, Dataton WATCHOUT control system, bespoke video content, aluminium-faced frame.

The Digital Wall
Ramus Illumination

Sydney, Australia

Opposite bottom | Shoppers visiting Sydney's Central Park retail centre are able to affect dynamic imagery playing across one of the world's largest (37 sq m; 400 ft²) internally installed LED media walls. *The Digital Wall* presents pre-designed video content that can be altered from a touchscreen kiosk or via visitors' motions captured on ceiling-mounted cameras.

Technology | AirLED-7 LED screen panels, VDWall LVP603S video processor, Crestron QM-RMC 2-series control system, SpinetiX HMP200 media player and SpinetiX Elementi X scheduling software, NEC V322 32" multi-touch screen, Audica MICRO sound system, Microsoft Kinect 1 depth camera, Arduino circuit board and custom lse sensor, 3 x computers running Windows 8 Pro, 4 x Panasonic WV-CP624 cameras.

Exhibits

Interior Intrigues

Radiant Geometry

Skylight II
Ljusarkitektur, Inaba

Stavanger, Norway

Inverting traditional designs of chandeliers, computer-controlled stripes of colour-changing LEDs activate the interior surfaces of *Skylight II*, a giant cylindrical light sculpture suspended in the glass-walled atrium of the five-storey New Concert Hall in Stavanger, Norway. Developed by New York design consultancy Inaba with Norway's KORO Public Art, the 'inverted chandelier' measures 11.5m long (37¾ ft), with a diameter of 6.6m (21½ ft), and weighs 6.5 tonnes. Aluminium tubes have been installed as gleaming stripes around the exterior of the Möbius-inspired sculpture, all aligned parallel to its long axis. *Skylight* produces distinct patterns and colours of light in response to different tones of the Nordic sky and different times of activity in the concert hall.

Technology | 662 x Philips Color Kinetics iColor Cove LEDs, Pharos control system.

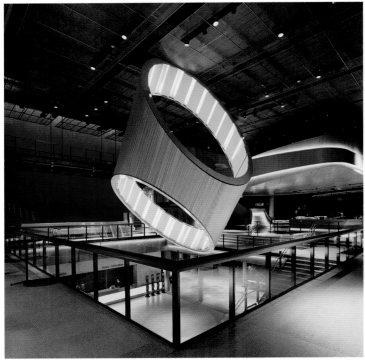

Exhibits

Interior Intrigues

Artificial Biologies

Hylozoic Structures
Philip Beesley

Various locations

Toronto-based architect and media arts and sciences professor Philip Beesley is dedicating his career to creating artificially breathing jungles. Inspired by hylozoism – the ancient spiritual-romantic idea that all matter lives – he uses advanced digital visualization, mechatronics engineering, robotic prototyping, artificial intelligence advances, and tiny sensor and LED light systems to craft extraordinarily intricate and sensor-responsive sculptures. Usually Beesley's arrangements of artificial vegetation are suspended in large gallery spaces, allowing visitors to (carefully) walk around and through the 'foliage', often triggering quivers and other delicate reactions.

Top and bottom left | *Epiphyte Chamber*, 2014, Seoul, Korea.
Opposite bottom left | *Protocell Field*, 2012, Rotterdam, The Netherlands.
Opposite bottom right | *Aurora*, 2013, Edmonton, Canada.

Technology | Sensors, various LEDs, thought-controlled electro-encephalograph headsets (*Protocell Field*), control systems.

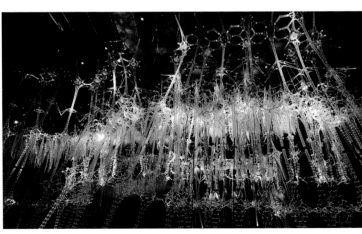

Exhibits

Interior Intrigues

Electric Thrills

Lumino-Kinetic Dramas
Bálint Bolygó

London, UK

Using LEDs and lasers with fine wire, mirrors, lenses and small electronic devices, London artist Bálint Bolygó plays with the laws of light reflection to generate mesmerizing spatial dynamics.

This page | *ArRay*: An octagonal prism splits the light ray from one LED into eight divergent shafts which reflect off subtly moving surfaces controlled with Nitinol 'smart memory' wire. Depending on ambient temperatures, the Nitinol nanostructure allows the metal to change shape at an atomic level, thus changing length and acting like muscle wire. This lumino-kinetic sculpture transmits magnified, mysterious images of these organic transitions.

Technology | 1 x LED, optics, stainless steel, Nitinol wire, electronics.

Opposite left | *Pulsar*: One strong laser light beam is manipulated to create a total internal reflection on a concave mirror and precise vibrating effects similar to those of pulsars (neutron stars).

Technology | 1 x 150mW laser diode, DC motor, aluminium, plywood, mirrors.

Opposite top right | *Sky Loupe*: This work fractures laser beams with optical gratings that cause almost random interference patterns. Ephemeral plays of coloured light keep viewers in constant suspense – perhaps not knowing that the light effects are powered by their own footsteps on Pavegen tiles, which harvest kinetic energy for off-grid power.

Technology | Aluminium, acrylic, laser diodes (50-500mW), motor, electronics.

Opposite bottom right | *Aurora*: Coloured laser beams – inspired by polar auroras – glide and bounce around a lobby ceiling at London's old Bethnal Green Town Hall (now a hotel). These effects were created with small rotating and fixed mirrors.

Technology | Laser diodes (100-150mW), mirrors, DC motors, asymmetric cyclic timers.

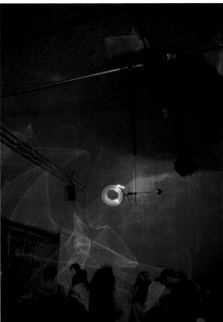

Exhibits

Interior Intrigues

New Babylonian Cities

Leisurescapes of Networked Light and Data

Vesna Petresin

Sector Jaune (Yellow Sector) concept by Constant Nieuwenhuys, 1958 (detail).

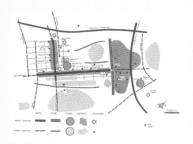

Visual form of Los Angeles as seen in the field, from *The Image of the City*, Kevin Lynch.

Fear of darkness is a primal instinct for humans. We need light to cast away shadows and defeat our anxieties about the unknown. Sight is the sense with which we recognize light reflecting off the surfaces around us, allowing us to understand and navigate our environment. As a survival tool, light allows us to 'tame' the dangerous night and fulfill our need for visibility, orientation and security.

From the industrial revolution to today's digital age, we have been shaping modern society with increasingly sophisticated ways of manipulating light. No longer merely a flickering medium for limited revelation (cave fires, candles, oil lamps), light is now conveyed with diverse instruments that can be orchestrated remotely and subtly, like (and often with) music, to stir and satisfy many senses and emotions.

Following the dramatic changes to civilization triggered by the discovery of electricity, we are now experiencing another paradigm shift: from artificial light as a means of visual perception, communication and navigation to light relayed as bits of digital information that can be diversely expressed across globally teleconnected devices and places.

Computer-controlled light is enabling unprecedented transformations of cities into datascapes, involving frequent deconstruction of static architectural façades to become immaterial screens transmitting kinetic and dynamic scenes.

We are the 'New Babylonians', suggested New York-based architecture critic Mark Wigley in a 1998 book that reinterpreted Dutch artist-architect Constant Nieuwenhuys's post-Second World War visions and models for a Utopian leisure city.

Our urban habitats are centrifuges of trade and competition. Being louder, larger, faster, brighter and more visible attracts commercial success and political power. However, aggressively filling the 'image of the city' (Lynch, 1960) with constant barrages of light and information still reveals our infantile fear of the dark and invisible; our trepidation in confronting void, error, ambiguity and instability.

Pauses and uncertainty are natural phases in the evolution of all species, including humans. But in today's Google era there is little respite from relentless influxes of data. Increasingly auto-tagged with xyz location co-ordinates, unique physical characteristics, and unique real and virtual behaviour patterns, information is streaming around cloud accounts and server farms in quantities far more than we know how (or why) to analyze and manage.

Global media culture tolerates neither silence nor darkness. As our cities grow, their expansions and densities are precisely marked with light. When Earth is scanned from a satellite or space station, every night is Christmas. How much light is too much?

Casting excessive artificial light, where and when it is not needed, interferes with human sleep patterns and visibility of the night sky, disrupts other biological processes, and wastes significant energy, material and financial resources. Light has been linked to diverse types of pollution in cities, and light pollution affects our health – psychologically, physically and economically.

All living creatures reproduce and evolve through incremental responses to Earth's natural cycles with the sun and moon. Our wellbeing depends on light to support our circadian clock of 24-hour biological rhythms synchronized with the Earth's day-night cycles. To balance our energies between activity and sleep, we require exposure to sufficient and correctly timed amounts of light, from appropriate bands of the electromagnetic spectrum.

We are gradually gaining more precise understandings of how light quantities, spectra, timing, duration and distribution affect our whole physiological system. One possible cure for light-related health disturbances may be programmable outdoor lighting combined with a facility for applying and measuring light that is effective to both visual and circadian systems. This concept is being investigated by the light and health program at the Light Research Center of the New York State Energy Research and Development Authority.

Current trends suggest we may live longer, and global population growth is still accelerating, but we also now recognize how seriously our economic growth – still mostly fuelled by Earth-extracted resources rather than renewable energy – is threatening the lives of future generations.

Energy efficiency does not necessarily mean energy reduction. For example, low-energy LED lamps are now proliferating at massive and exponential rates of production, obviously threatening to increase substantially rather than (as promised by manufacturers) reduce our planet's total consumption of power. 'Sustainability' may be a concept misplaced in advertising and designs promoting large arrays of LEDs. OLEDs, using organic compounds, are one step forward to possible biological improvements.

Along with recent advances in internet-related technologies to enable smart cities, research in quantum physics, chronobiology and neuroscience may provide new understandings of nature-related solutions to contemporary urban challenges (Armstrong, 2015).

In addition, multimedia art techniques and applications may provide a new language and inspirations for developing environments that integrate both material structures and immaterial flows of time, information and social behaviours.

Across all urban environments throughout history, light has delivered major social, economic and aesthetic impacts. In a 2010 MIT thesis titled *Liberating Pixels: Alternative Narratives for Lighting Future Cities*, Dr Susanne Seitinger clarified that lighting displays evolved through complex interleavings of the social and the material towards programmability, addressability, responsiveness, mobility and ad-hoc control.

 Pixels can be deployed flexibly within any environment or scale for aesthetic purposes, including visualizing a city's data for enhanced navigation and signalling. Applications of urban visual communication have been expanded through wider availability of LED lamps, along with advances of miniaturized and embedded computation, such as sensor-responsive LED systems.

Hypersurface theory anticipated merging of the material and the immaterial by resisting binary oppositions such as structure/ornament. 'Hypersurface' is defined as the equivocal interplay between media and material such that neither format is fully identifiable. Thus, hypersurface architecture involves both the fluidity of information and the malleability of form or material, combined as one integrated and emergent design strategy.

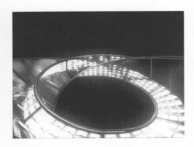

Double Möbius Strip, Laguna Beach, 2000-1, by Vesna Petresin-Robert and Laurent-Paul Robert, with Stephen Perrella.

Songdo International Business District at Incheon, South Korea.

Buildings still have supporting structures and enclosing surfaces. However, with the emerging role of hyperminiaturization, wirelessness, digitization and dematerialization, the architectural functions of shelter have shifted to symbolic. In the city of pixels, digital ornament is enabled by the façade as interface.

Is information becoming a contemporary means of decoration? Robert Venturi (1996) reflected on substitutions of traditional forms of architectural ornament and symbolism for digital decoration, noting: 'The sparkle of pixels can parallel the sparkle of tesserae and LEDs can become the mosaics of today.' He imagined 'architecture as iconographic representation emitting electronic imagery from its surfaces day and night rather than architecture as abstract form reflecting light from its surfaces only in the day...'

Habitation involves continuous exchanges of information between a building and its occupants, and the introduction of electronics requires us to rethink these interchanges. Any part of a building can now be embedded with sensors, and any dynamic element can be controlled by a computer. These elements can be networked together and programmed, and we may read displays and operate controls or just up/download bits between the building net and our body net. Differences between the computer network and the structure dissolve. Interface becomes architecture, and architecture becomes interface.

The maxim of a 'liveable city' is more than a vision. Cities are implementing many health-promoting concepts in their designs and controls for both physical and information infrastructure.

In scholarly contexts, William J. Mitchell's 1990s 'smart cities' prophecies may be understood as an evolution of Bernard Tschumi's 1970s concept of an 'event-space'. While both these ideas are still generating diverse interpretations, some urban areas (such as London's South Kensington, South Korea's Incheon technology city and Singapore's Marina Bay waterfront renewal area) have understood their potentials for place-making by transforming tired and underperforming sites into net-enhanced environments with computer infrastructures that capture data from device users, support virtual interactions, and auto-push public or commercial information.

Setting up a 'smart city' generally includes provision of energy-efficient, user-responsive lighting, with systems that can analyze consumers' and residents' behaviour patterns. LED street lighting that captures visual and atmospheric data from the environment may provide new content and scenarios for a variety of events and uses of public space, perhaps enabling cost savings by connecting lighting to the internet, and deriving even greater value by using the lighting network for other connected services.

Intelligent, networked, public lighting infrastructures can enhance urban liveability by creating an energy- and resource-efficient, environmentally resilient city; by increasing traffic safety and personal security; by reinvigorating public spaces; and by constructing a vibrant city brand to communicate identity and enhance economic development.

While artists, designers and architects have been exploring the potentials for light art to expand from building façades to all urban surfaces, smart lighting infrastructure systems are now imperative in planning all major urban developments. Today, most cities are becoming smarter simply by transitioning from analogue to digital energy control systems.

Digital Transformations

Digital light augments cities, animates landscapes and expands the territories of light/data to include any surface or material support. Backed by pilot studies demonstrating that city leaders prefer LED lighting due to social and environmental benefits (such as improved safety and visibility), global electronics group Philips has claimed that switching to LED technologies could save cities 50 to 70% of energy (or up to 80% if equipment changes are supported by smart controls).

In London, more than 35,000 streetlights will be converted to LED lamps controlled by a central system – part of a government promise to bring lighting on the city's roads up to 21st-century standards in order to increase traffic safety and to lower carbon emissions. Equipment suppliers estimate potential power cost savings at £1.85 million a year.

The Berlin pilot project 'Clean Light City' includes WiFi-enabled 'light maps' visualizing real-time data captured by streetlights, and feedback of activities among users across the area. Potential services include programmable light with data captures supporting retail, urban navigation, guiding and path-finding systems.

Similar transitions from analogue to LED lighting, and other sustainable urban lighting strategies, are the subject of guidelines in 'Lighting the Cities', a recent European Commission report for municipal agencies. Case studies from twelve international cities are explained in *The Social Dimensions of Light*, a publication from the Urban Strategies Commission of Lyon-headquartered Lighting Urban Community International (LUCI).

With smart LED lighting and real-time data systems, municipal authorities such as transport providers can use lighting to improve passenger safety and comfort, reduce energy consumption and add new aesthetic values to urban environments. One study (Seitinger et al, 2012) suggested that LED accent lighting may contribute to more efficient transport by supporting flows of pedestrians through a public transit network without distracting them, while also creating richer, more informative experiences.

Future developments of street lighting will allow us to rethink ways in which we interact with our urban surroundings, and how and to what degree the city interacts with us. One emerging experiment is *If Light Could Fly*, a demonstration of LEDs fixed to drones flying along the Strijp-S at Eindhoven for the 'GLOW' light festival in 2013. The concept team – interaction designer Wolfgang Klein with Metatronics and Creative Innovation Works – suggest these remote-controlled, unmanned aircraft could 'nest' in clusters on city walls, then be activated by pedestrians to perform various useful outdoor lighting tasks.

Beyond lamp posts, digital lighting extends to all urban surfaces: façades, roads, water, roofs, landscape features and street furniture. For example, Philips's ArchiPoint technology uses sensors to light concrete pavements to appear translucent and luminous, while walls of LEDs, controlled by computer, display moving images. Any wall, building, bridge or paved surface could become a medium for communication, enabling unprecedented design and communication possibilities.

Experts at MIT and Philips have been developing other exciting innovations, including interactive shopfronts and ways to exploit decorative lighting for emergencies. They are also connecting light controls to other intelligent systems (for traffic, parking, water supply) via ubiquitous networks on light fixtures, with integrated sensors capturing real-time flows of environmental and user data.

If Light Could Fly, a 2013 drone performance at Eindhoven's Strijp-S park, by Wolfgang Klein with Metatronics and Creative Innovation Works.

The MIT-Philips Fluid Interfaces team, led by Susanne Seitinger, also introduced LightBridge weather-adjustable public lighting and Urban Pixels, a wireless network of individual, autonomous, physical pixels applicable to any urban surface. Another experiment, Augmented Reality (AR) Street Light, uses Boston's infrastructure networks to enable users to interact in real time with smart street furniture that is linked to LED lighting and QR codes (via cloud-based mobile apps).

In Britain, 'Starpath' coatings of luminous blue particles have been sprayed onto damaged tarseal pathways across the Christ's Pieces park in Cambridge, for a fraction of the cost of standard park lighting. Developed by Pro-Teq Industries, this material allows storage and absorption of ultraviolet rays during daytime, to be released as light after dark. While this may work well in areas with low budgets and difficult access, including open spaces or cycleways, it might also cause disruptive impacts on natural systems and habitats.

Street lighting is not only becoming programmable in terms of changing colour, orientation and luminescence (depending on usage and other issues); it will also allow new sensory interactions, with light being activated by breath, sound, heat or movement.

Networked Nomadicity

Environmental awareness, symbiosis with nature and technological development may not be mutually exclusive. Principles of emergent organization in nature can be transferred to the design of smart urban ecosystems. Cities can become smarter by optimizing their infrastructure. While using LED lamps may not really help us lower global energy consumption and costs, that ideal remains attractive, particularly at the scale of citywide networks of public lighting.

Today's adaptive, interoperable urban lighting systems are increasingly networked with other services through 'energy Internet' or 'smart grid' infrastructures. New energy grids could become two-way communications networks, allowing businesses and consumers to generate, store, sell and consume electricity. They could also serve as IP platforms, connecting individuals and devices, and innovating uses of data, communications and connectivity through open data flows between traditionally closed or proprietary infrastructure (for example, public lighting and traffic systems).

Plans for smarter cities should deliver operational savings from multiple services and platforms, exploiting existing lighting infrastructure with video, sensor and hotspot equipment to provide advanced visual communications for safety, traffic management and other government priorities.

In ideal terms, smart city lighting solutions are engineered to cater for the diverse needs of users, to reduce energy consumption and light pollution, and to support efficient management and repairs of equipment. Smart street poles, connected to a smart power grid, can enable many functions, from capturing environmental data (temperature, humidity, traffic density) to acting as electricity distribution points and WiFi hotspots. Energy networks now allow data and energy convergence, through new transfer and management protocols, the smart grid and smart loads requirements, and other environmental protection policies.

Coming from the commercial sector is 'Light as a Service' (LaaS), a business model for large equipment vendors to provide digital urban lighting networks to city authorities and energy utilities. LaaS providers promote integrations of data streams, network connections and energy management systems, while extending control capabilities for consumer devices such as smartphones and tablets. Auto-responsive lighting controls are being integrated with management systems for buildings and public spaces to help optimize energy consumption and user interactions.

Contemporary public lighting infrastructure operates as a 'network of networks' and provides a platform for service innovation. This is generating a new form of urban nomadicity, enabled by seamless connectivity. Users today require immediate and constant access to ever-increasing quantities of audio-visual material.

By selectively loosening place-to-place continuity requirements, wired networks have been producing fragmentation and recombination of familiar building types and urban patterns. For example, bank branches have been replaced or at least supplemented by cash machines, and location-based advertising is now combined with electronic urban navigation.

Gradually emerging from the messy but irresistible implications of ubiquitous wireless coverage is the possibility of a radically reimagined, reconstructed electronic form of nomadicity, grounded in a well integrated wireless infrastructure, combined with other networks and deployed on a global scale.

Network connections are fluid and amorphous. Although contained by physical cables, electronic impulses are not structurally rigid like buildings. Some public services that were delivered with architecture, furniture and other urban fixtures are now being provided from implanted, wearable and mobile devices. Activities that once depended upon close proximity to sites of accumulation now rely increasingly on mobile and global network connectivity.

Democratizing Light

Today's planning of data cities and Light as a Service could help deliver the 'Non-Plan' vision promoted by Reyner Banham, Paul Barker, Peter Hall and Cedric Price in a 1969 issue of the social affairs magazine *New Society*. Aligning with French philosopher Henri Lefebvre's insistence on 'the right to the city', they claimed that the most daring development plans for cities (as with Hausmann and Napoleon III's interventions across Paris) usually produce the least democratic places to occupy. As an antidote to relentless impulses to build 'megastructures', they proposed new strategies for 'plug-ins' (services) and disposable architectural elements that could be 'democratically' configured by occupants.

More than fifty years later, today's citizens of prosperous precincts are regularly using automatic services such as electronic road charges, satellite-linked navigation maps, and convenient combinations of electronic monitoring of parking space occupancy and automatic direction to vacant spaces.

As well as these useful new examples of public and mobile transactions of data, new genres of public art are being developed by users of apps on mobile devices, who are remotely controlling data flows and colour and light effects in outdoor spaces.

One notable expression of neo-democratic light art was *Water Light Graffiti*, a 2012 production by French artist Antonin Fourneau with Digitalarti's Artlab in Paris. Citizens used water to activate LED screens: an experiment that redefined the public painting specialties of frescoes and urban graffiti, integrating them as one different genre – user-generated electronic illustrations.

An obvious concern about encouraging crowds of people to generate their own light art is the potential for this creatively exciting concept to increase global consumption of energy exponentially. Power is still delivered mostly from Earth-dug fossil fuels that are disrupting familiar patterns of environmental behaviour at all geographic scales.

Will light become a scarce luxury in future? Certainly the ideal vision of democratized light is clouded now by the widely recognized need for humans to take personal responsibility for the amounts of energy they consume – at least until humanity can depend extensively on renewable fuel supplies. This concept has gained momentum globally through the annual 'Earth Hour "switch off your lights"' promotions developed since 2008 by the World Wildlife Fund for Nature.

One energy-reduction strategy is for communities to source their own power supplies, reducing their reliance on national grids. This concept is being pre-empted by many governments selling either publicly owned 'poles and wires' or energy supply licenses to commercial operators.

Harnessing Nature To Evolve Technology

Leading researchers suggest exploiting existing natural light phenomena to help reduce our energy consumption or bypass our dependence on energy generation. Examples include phosphorescence, light-emitting gases, bioluminescence, moonlight and auroras such as the Northern lights.

Water Light Graffiti demonstration in Paris of water being sprayed to activate a wall of LEDs, by Antonin Fourneau and Digitalarti Artlab (2012).

London studio Random International's *Lumiblade* interactive wall of Philips OLED tiles (2009).

American scientist Buckminster Fuller proposed a worldwide revolution led by 'comprehensive designers' who would globally coordinate resources and technology for the benefit of all mankind, anticipating future needs while implementing resource-consciousness across all aspects of planning and design.

Today, many designers are proposing customized products embedded with bio- and nanotechnologies. Barriers between organic and mechanical processes are slowly eroding. This trend also affects designs for light fixtures, services and visual effects, increasingly including biomimetic (nature-inspired) solutions with user options for multi-tasking.

Light sources will soon appear – and operate – very differently from last century's light bulbs. Contemporary researchers are developing new materials and forms of illumination, including non-woven sheets, and flexible and water-resistant luminaires and screens, as well as fixtures printed in 3D or 4D.

Architectural applications range from OLED tiles (where light is emitted from a layer of an organic compound), luminous paving blocks and LED-wave guiding systems to glowing street furniture and illuminated landscape features such as bioluminescent trees. Sensor-responsive, low-resolution LED screens can be used for both urban lighting and displays, while, at the personal scale, handheld LED lights, along with wearable, implanted, mobile devices and nano-tech eyeware, are becoming prevalent. Solid state hardware for communications and computing frequently now includes AMOLED (active matrix organic light emitting diode) screens, while some companies and inventors are printing electronics as components for luminaires.

With rapid escalation of the internet of things (IoT) phenomenon (computerized products exchanging situation data and action messages across wireless networks), lights are being linked in clusters that behave in apparently spontaneous yet obviously communicative patterns. Moves by one or a few of the group will trigger rapid responses from all participants, like flocks of birds or swarms of insects.

IoT-linked lighting devices are also being developed to respond to users' moods and even help look after users' health and wellbeing. When hand-held or worn, and using purpose-coded apps, these can detect emotions and change colours to stimulate different moods. As well as providing light, some devices and apps now provide detailed updates of personal information on health-related conditions, with suggestions and options for improvement.

Emergent Urbanism

At global technology companies, researchers and innovation engineers are developing smart city systems using scientific concepts of emergence (where individually simple entities co-operate to form complex and unpredictable systems and phenomena).

It seems increasingly relevant to apply emergence and other systems principles from particle physics to urban datascapes. Observations of self-organizing, complex and non-linear, non-hierarchical systems, distributed networks and particle fields are rich sources of new design ideas. These fields of knowledge have potential to inspire urban professionals to plan built environments that are more responsive to specific, changing needs and desires among individual users.

To deal with change as a basic condition when designing places for human life, scientific observations of transformation processes can help inspire new strategies. In theories of open systems, for example, structures evolve through transitory states at all scales of the system. They are influenced and transformed by force fields and many other external factors; they adapt to feedback and internally reorganize in anticipation of change.

In urban planning, self-regulating orders and patterns (growing into complex networks) can incorporate variations of time as a field condition, without destroying the internal coherence of the prevailing system. In such cases, variations become localized, boundaries shift and 'hierarchies' are fluid.

Data can be dynamically visualized using animation and interactivity. Most types of representation can be mapped onto, or morphed with, one or more types: numbers into images, images with sounds, sounds pervading spaces – all processes closely relevant to urban lighting design and delivery. New visual representations often initiate not only new paradigms, but also transformations of design processes and modes of production.

Enabled by the data-crunching powers of computers, emergence and many other theories of complex dynamic systems are now essential to analyze forms and behaviours of both physical urban environments and online social networks. As ever, the most sophisticated scientific interpretations of human behaviours are initiated by the defence and public security agencies of the world's richest governments.

During our present phase of species evolution, humans seem to be increasing their cognitive capacity for sensitive intuition and complex reasoning. Inherently we are tempted to heighten and extend our sense experiences, and these primal impulses can increasingly be satisfied via mobile, always on, multimedia technologies.

Individual behaviours (in physical and/or virtual realms) influence whole urban systems through feedback and growth, causing interdependent development. Every event exists in relation to surrounding events and entities that constitute the field.

Individual Behaviours in Physical or Virtual Cities

Urban environments are no longer shaped merely by boundaries but increasingly by network connectivity. And just as technology does not necessarily develop as its designer envisaged, so citizens – not only architects and planners – may increasingly determine the functional and symbolic aspects of space.

However, smart urban environments cannot rely on or be developed only with technology. Beyond mechanistic processes and services, societies always depend on human exchanges of empathy and mutual understanding of every individual's duty to support others.

Light has become today's most exciting tool and platform of urban creative expression, as architecture, engineering and construction professionals evolve their goals from producing static compositions of inert building materials to dynamic interactive arrays of components and their interfaces.

Day and night, city structures are operated and mediated by sensors, digital controls and feedback mechanisms that are programmed to anticipate and adjust to evolving patterns of occupation. Conversely, this century's vast potentials to creatively transform night environments with new expressions of light require us to rethink our human values of individual responsibility, co-operation and symbiosis with other species and natural phenomena.

Illuminating the Titanic

Or Changing Course?

Conclusion
Peter Droege

This book presents inspiring ideas and initiatives for 'eco-ethics' in urban illumination. Given our escalating global exposures to extreme and man-made environmental crises, these 'sustainable' concepts could amount to little more than lighting decks and dining rooms on the *Titanic* – unless combined with urgent and substantial moves away from fossil and nuclear energy systems and towards a 100% renewable world.

Our most hopeful (and indeed only feasible) vision for empowering cityscapes surging with endless streams of light and data is to fuel all of their operations with renewable energy.

Along with energy-efficient building designs, renewable sources – wind, water, solar, solar thermal, geothermal and bio-energy – are essential to support leaner, healthier, more prosperous and more equitable civilizations.

Unlike fossil fuels, renewable sources provide a long chain of benefits. They reduce pollution of local and global environments, secure enhanced levels of prosperity more equitably, slow depletion of natural resources such as fresh water and minerals, halt (and potentially help reverse) catastrophic declines in biodiversity, and improve human health.

Only renewable fuels can help avert nuclear explosions (by rendering obsolete this lethal and costly energy system), replace peaking supplies of conventional fuels and increase the resilience of urban areas facing climate change disasters.

Across this planet, all human energy needs can be met by supplies of locally appropriate combinations of renewable fuels, if consumer demand and delivery efficiency are intelligently managed.

For every city and its surrounding region, independent energy islands can coexist with smart power grids and distributed storage systems – mechanical, electric, chemical or thermal. Such energy storage can be introduced as fixed infrastructure or can be mobile (vehicle-based). Many successes in local and regional renewable energy autonomy already demonstrate how renewables solve the major urban challenges faced by communities and governments.

Given the right policies, renewable energy systems can be introduced incrementally and generate local income. Many studies have demonstrated this potential over the past three decades, from France's 1978 ALTER report to Mark Z. Jacobson's team research at Stanford University in 2009.

Infrastructure of the fossil fuels era: motorway spaghetti in Shanghai (photo SevArt/PicFair).

Conclusion

Illuminating the Titanic

Carbon-belching fossil fuels supply 85% of global commercial energy, with renewable power sources at 12.5%.

City lighting is the most excessive use of power, and too much light pollutes citizens' views of stars in the sky.

Energy Systems are Social Systems

Energy systems are not merely the technical infrastructures to deliver heat, power and mobility. Across industrial countries and the developing world, energy systems are inseparable from socio-economic and governance structures and interests.

All the main benefits of renewable energy are widely documented and accepted in principle, but powerful institutions and industries have been shaped around – and continue to protect – dangerously outmoded business models. Economic worlds and social relations are moulded (and trapped) by the realities of centralized systems based on extracting, processing and distributing oil, coal and uranium, and converting and selling its energy.

While hundreds of thousands of rural Bangladeshi villagers have benefitted from Grameen Shakti's microloans for installing solar power (a system in principle equally applicable to informal urban settlements), millions more people are miserably trapped either far from power supplies or in urban slums, where they are required to fight for scarce and costly cans of kerosene.

Because renewable energy allows cities to generate and manage their own local power supplies, communities now have potential to capture the enormous funds they currently spend to support oil, coal and uranium extraction and processing in faraway places.

By investing these massive streams of consumer expenditure into local systems providing renewable power, cities go beyond reducing their energy demand through efficiency and conservation. This model of productivity is essential to secure the future viability of cities.

Cities are not autonomous worlds: no city is an island. They do not depend only on globalized trade (cherished widely as the putative bulwark of international stability and prosperity); they also depend hugely on nearby regional and rural areas for resilience in times of need. Even more profoundly, cities depend on the now rapidly shrinking domains of undisturbed forests, wetlands and waters, sources of oxygen and sinks for tragically increasing streams of industrial waste, including nitrogen, ammonia, methane and carbon dioxide.

Globalized flows of trade and communications have caused cities to become detached from their hinterlands, while global resource stream and waste assimilation sinks are now at the verge of collapse. The building of regenerative regional and circular resource systems is an inexorable and exciting challenge.

Cities need bold strategies for ecological and economic resilience and permanently reliable supplies of resources. But the required responses are not being implemented. Why not? Political ignorance and concerted campaigns by fossil and nuclear fuel-based industries and institutions have undermined public policy moves repeatedly over the past forty years, including the systematic demolition of national and local rail services to encourage expansive building of road networks for privately owned cars. In addition, many environmental protection groups have ironically failed to integrate accelerating supplies of renewable energy in conservation schemes, deflecting public attention instead to fear tactics, deckchair shuffling and largely symbolic 'sustainability' strategies, such as selectively counting and cutting carbon emissions, and exhorting citizens to reduce water consumption, swap light bulbs and shopping bags, and recycle household waste.

Many policymakers are confident that there is still time for successful action but less confident on how to expect a dramatic shift in the face of intransigence among the incumbent energy supply and policy institutions – and resultant standoffs over many years of United Nations-managed global climate negotiations. Even the massively ambitious Global Earth Observation System of Systems (GEOSS) and Digital Earth science communities are focused mainly on monitoring and modelling environmental solutions. They are not directly accelerating the essential shift from wrong to right sources of powering the exaflops and yottaflops (future computer-processing speeds of bits per second) that will be required to run their envisaged global spatial data infrastructure.

Proposing a *SuperLux* Code of Energy Conduct

How does our world's urgent need to adopt renewable energy specifically relate to light artists and designers? Are lighting professionals part of the problem or solution? Villains or saints?

Light design and art is wonderful – but also a form of conspicuous consumption that often does not reflect concerns about the fossil or nuclear content of these displays of creativity. Yet designers, artists and cultural producers ride on the visual cortex of human civilization, directly stimulating its sense of reality: hence, they are expected to act responsibly, as leaders. Much of what we see at night is illuminated by 'artificial' light. Let it all be solar and renewable in origin!

'Smart Light Sydney' was, in 2009, the world's first light art exhibition that claimed eco-positive energy consumption, via offsets. Offset action is a useful start, but only a start. Without replacing all fossil and nuclear content from the global energy system, carbon offsets cannot be effective in answering climate change and conventional energy risks. But together with a complete conversion, the right kind of offsets – notably the sustainable and sustained biosequestration of atmospheric carbon – are crucial measures to be deployed in synchronization with phasing in renewables.

In concluding this *SuperLux* global survey of outstanding works of urban light art, it seems essential to call on all artists and designers to align with a new code of conduct to strengthen humanity's future in our new and delicate millennium. Let us all decide to design and commission only light systems that are powered renewably – without fossil or nuclear energy content.

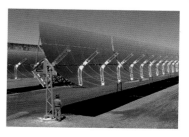

Large arrays of solar thermal collectors are humanity's best future source of renewable power.

Melting ice in permafrost areas such as Alaska is catastrophically collapsing paved roads and buildings.

City Lights from Space

Astronaut Maps
of Urban Life at Night

Nightpod Images from ESA/NASA

Patterns of streetlights and densities of
building lights show astronauts in space
where humans are living on Earth. Colours
on these urban lighting snapshots indicate
different technologies prevalent in different
places. To help capture Earth images in sharp
focus, the European Space Agency developed
a Nightpod camera tripod with a computer-
assisted 'nodding mechanism'. Set up in the
Cupola of the International Space Station,
it allows fixed digital cameras to reconcile
discrepancies between the Earth's speed
and track around the sun and the ISS's orbit
around the Earth.

This page (from top) | Paris,
San Francisco Bay Area and Berlin.
Opposite page: Ciudad Juárez and El Paso.
Cities not shown at same scale.
All images copyright ESA/NASA.

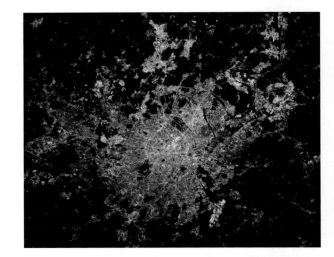

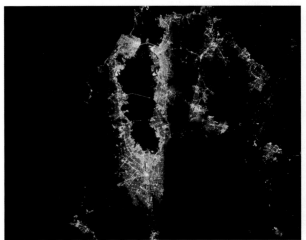

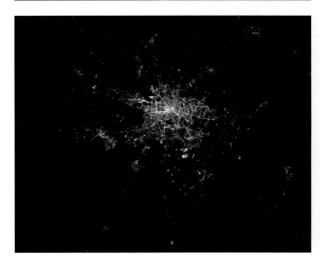

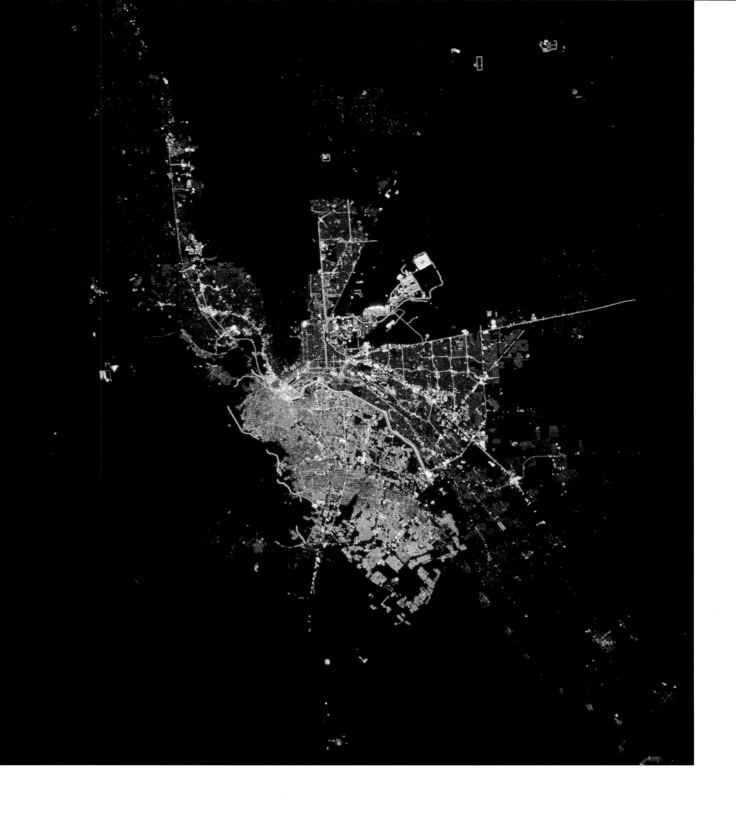

City Lights from Space

City Light Festivals

Light festivals date back to artistic arrays of torches in ancient cities of Asia and the eastern Mediterranean. Today these celebrations are globally diverse in their themes, scales, durations, art curatorial strategies, local popularity and international prestige. Here is our list of notable festivals, which should be cross-checked online for dates, frequencies and cancellations. Germany, the Netherlands and France are reputed to host the most innovative light events. Also check cities launching new light festivals.

International

Diwali (aka Deepawali) Ancient international Hindu light celebration, with similar festivals by Jains and Sikhs: lunar date from mid October to mid November (annual)
Earth Hour One-hour 'switch lights off' ritual promoting reduced fossil fuel consumption: last Saturday March (annual, first year 2008)

Europe

Alingsås Lights in Alingsås, late September-early November (annual)
Amsterdam Amsterdam Light Festival, late November-January (annual)
Berlin Berlin Festival of Lights, mid October (annual)
Blackpool Blackpool Illuminations, August-November (annual, first year 1879)
Brussels Festival des lumières, early November (annual)
Cascais Lumina, mid September (annual)
Chartres Chartres en lumières, April-November (annual)
Copenhagen Tivoli Christmas Market, December (annual)
Desio Kernel, late September (annual, since 2011)
Durham Lumiere, mid November (annual)
Eindhoven GLOW, mid November (annual)
Frankfurt Luminale, early April (biennial, since 2000)
Gdansk Narracje, mid November (annual)
Ghent Light Festival Ghent, late January (triennial)
Hamburg Lux Hamburg, mid February (first year 2015)
Helsinki Lux Helsinki, early January (annual)
Leipzig Lichtfest Leipzig, 9 October (annual, since 2009)
Ljubljana Lighting Guerrilla, May-June (annual)
London Regent Street Christmas Lights, November-December (annual)
Lüdenscheid LichtRouten, September (quadrennial)
Lyon Fête des lumières, early December (annual, first year 1852)
Paris Champs Elysées Christmas Lights, December (annual)
Prague Signal, mid October (first year 2013)
Tallinn Valgus Biennial, late November (biennial)
Toruń Bella Skyway, late August (annual, first year 2009)
Turin Luci d'Artista, early November (annual)
Uppsala Festival of Lights, November (biennial, first year 2008)
York Illuminating York, late October-early November (annual)

Asia

Beijing Switch On Beijing, mid August (annual)
Goa The Story of Light, mid January (first year 2015)
Hiroshima Dreamination, November-January (annual)
Hong Kong A Symphony of Lights (nightly)
Kobe Luminarie, December (annual)
Osaka Hikari Renaissance, December (annual)
Seoul Seoul Lantern Festival, November (annual, first year 2009)
Singapore iLight Marina Bay, March (biennial); Orchard Road Christmas Light-Up, December (annual)
Tokyo Christmas Lights, December (annual)

North America

Chicago Magnificent Mile Lights Festival, November-December (annual)
Hawaii Honolulu City Lights, December (annual); Kaua'i Festival of Lights, December (annual)
Marshall Wonderland of Lights (Texas), December (annual)
Montreal Montréal en lumière (including Nuit Blanche), late February-early March (annual)
New York New York Festival of Light, October-November (first year 2013); Christmas Lights, December (annual)
Owen Owen Sound Festival of Northern Lights (Ontario), mid November-January (annual)
Sarnia Celebration of Lights (Ontario), December-January (annual)
Toronto Cavalcade of Lights, late November (annual)
Vancouver VanDusen Botanical Garden Festival of Lights, December (annual)

South America

Medellín Medellín Christmas Lights, December (annual)
Mexico City FILUX (Festival Internacional de las Luces), early November (first year 2013)

Australasia

Auckland Auckland Lantern Festival, mid February (annual); Diwali Festival of Lights, October (first year 2014)
Canberra Enlighten, late February-early March (annual)
Hobart Dark Mofo, June (first year 2013)
Melbourne White Night Melbourne, late February (annual, one night); The Light in Winter, June (annual); Gertrude Street Projection Festival, July (annual)
Sydney Vivid Sydney, late May-mid June (annual)
Wellington Wellington LUX Winter Festival, June (first year 2011)

Middle East

Dubai Festival of Lights, late March (first year 2014)
Jerusalem Light in Jerusalem, mid June (annual)

With thanks to Bettina Pelz for her advice as moderator for this international network of light art festivals.

About the Authors

Prof. DI MAAS **Peter Droege** is General Chairman of the World Council for Renewable Energy and holds the Chair of Sustainable Spatial Development at the University of Liechtenstein. During three decades of practice, research and teaching at MIT, Tokyo and Sydney universities, he has won many international prizes for urban design and sustainable energy advocacy. He has published historically significant books and many articles developing and promoting advanced urban planning and regenerative design principles, encouraging the practical replacement of fossil and nuclear sources with renewable energy, while regenerating land and water systems to sequester atmospheric carbon.

Davina Jackson is a Sydney-based international writer and promoter of cultural advances across design, architecture and digital arts. A visiting research fellow with Computing at Goldsmiths, University of London, she was a director of government-funded companies which produced the world's first 'Smart Light' festivals in Sydney and Singapore, and has curated exhibitions and published internationally on 'data cities', 'virtual nations', 'Digital Earth' and other geospatial policy innovations. A former editor of *Architecture Australia*, multi-disciplinary design professor with the University of New South Wales and author/editor of numerous books, blogs, newspapers and design magazines, she gained an M.Arch on digital home theories and is now publishing her PhD history thesis.

Prof. **Mary-Anne Kyriakou** chairs lighting design at the University of Applied Sciences Ost-Westfalen Lippe in Detmold, Germany. She created and directed the world's first eco-ethical 'Smart Light' festivals in Sydney ('Vivid'), 2009, and Singapore ('iLight Marina Bay'), 2010 and 2012. A founding leader of Meinhardt, the light art group for engineers, she was named one of the world's top 100 leaders of sustainability advances by ABC Carbon Asia in 2012. She studied electrical engineering, lighting design and music composition at the University of Sydney, has won awards for both design and music, and is a director of the Studio Kybra practice with German light architect Ingo Bracke. After patenting a brainwave-activated lighting control device, she continues post-Masters research on potential urban applications of both biological and technological luminescence factors.

Dr **Vesna Petresin** is a time architect, space composer and performer. Currently a visiting research fellow with Goldsmiths College, University of London, and an artist-in-residence with the ZKM Center for Art and Media in Karlsruhe, she earned her PhD for research on temporal composition in architecture, art and music. In 2004, she joined French artist Laurent-Paul Robert to establish the London-based Rubedo art collective, which integrates sophisticated themes and techniques from optics, acoustics, complex geometry, psychology and synaesthesia. The Rubedo team has delivered performances, immersive experiences, multimedia installations and artefacts to international festivals and venues, including the Vienna Secession, Tate Modern, ArtBasel Miami, Venice and Beijing Architecture Biennales, Royal Festival Hall and Sydney Opera House.

Dr.-Ing. **Thomas Schielke** studied architecture at the University of Technology in Darmstadt, Germany, and now leads lighting seminars and workshops at the DIAL Academy in Lüdenscheid. From 2001 to 2014 he worked at the lighting manufacturer ERCO, where he designed an extensive online guide for architectural lighting and led lighting workshops. He is a co-author of the ERCO book *Light Perspectives: Between Culture and Technology*, has published numerous articles on lighting design and technology, and has lectured at leading European and American universities, including Harvard GSD, MIT, Columbia GSAPP and ETHZ.

Prof. **Peter Weibel** is Chairman and CEO of the ZKM Center for Art and Media in Karlsruhe, Germany, and has been awarded Germany, France and Austria's highest honours for arts and letters. Educated in literature, medicine, logic, philosophy and film in Paris and Vienna, he recently received two honorary doctorates from universities in Helsinki and Pécs, and has held professorial positions with leading European and American universities. Since his 1986-1995 term as artistic director of the Ars Electronica Center in Linz, Austria, he has curated many milestone exhibitions, including 'Light Art from Artificial Light' at ZKM in 2005-6 and the 2015 'lichtsicht 5 Projection Biennale' in Bad Rothenfelde.

Project Credits

Il Giardino Verticale, Turin, Italy p.001

Technology 78 x Ilti Luce Z-Power RGB 5W LED lamps, 180 x power packs (tree), 246 x Ilti Luce LED lamps (flower boxes), 230m (755 ft) Ilti Luce Strip Riga RGB LED strips. **Elements** 520kg (1,150 lb) galvanized iron, 590kg (1,300 lb) stainless steel pipes. **Applications** Illuminated artificial landscape/sculptural installation. **Appearance** 'Heavenly' garden in a formal courtyard paved with a Baroque-inspired pattern using resin bars and LED strip lighting. **Completed** 2013 (Luci d'Artista festival). **Host/Venue** The Number 6/Palazzo Valperga Galleani. **Artist** Richi Ferrero (Gran Teatro Urbino). **Developers** Building Engineering. **Collaborators** Alessandro Amici, Valentina Gamba and Flavio Pollano. **Technical Consultants** Giuseppe & C. Bossi, Comor (building models and resins), Epaini, Frea & Frea, Ilti Luce, Tekspan. **Image** Marino Ravani. **Online** giardinobarocco.blogspot.it, richiferrero.it

Giant Fish, Hong Kong, China p.004

Technology 250 x Traxon Dot XL-6 RGB LED fittings, e:cue e:pix VMCs, bespoke software. **Elements** 2,500 x Chinese paper lanterns, timber and bamboo structure. **Applications** Temporary festival artwork. **Appearance** Giant colour-changing flying fish. **Completed** 2011 (Lee Kum Kee Lantern Wonderland celebration). **Venue** Victoria Park. **Producer** Hong Kong Tourism Board. **Architect** CL3. **Lighting Consultant** LEDARTIST. **Equipment** Traxon, e:cue. **Image** courtesy LEDARTIST. **Online** cl3.com, ledartist.com

Pixel Cloud, Reykjavik, Iceland pp.006–007

Technology 2 x Panasonic PT-DS12KU 24K lumens projectors, AKAI MPD26 midi controller, MadMapper and Modul8 software. **Elements** Steel scaffolding, white net fabric. **Applications** Temporary outdoor multimedia performance. **Appearance** 'Cloud' of translucent white cubes by day, transformed with projection-mapped digital imagery at night. **Completed** 2013 (Winter Lights festival). **Organizers/Sponsors** Höfuðborgarstofa, Orkusalan, Iceland Design Centre. **Venue** Austurvöllur Square. **Artist** UNSTABLE (Marcos Zotes). **Music** *Cosmos* by Eðvarð Egilsson. **Technical Production** Luxor. **Project Management** Gerður Sveinsdóttir. **Scaffolding Construction** ROST. **Fabric Installation** Marcos Zotes, Erika Barth (Veraldavinir), Edda Guðmundsdóttir (Shadow Creatures) and LHÍ students. **Videography** Brandenburg, Jóel Sigurðsson. **Image** Marcos Zotes. **Online** unstablespace.com

Water Cube, Beijing, China pp.010–011

Technology 440,000 x Cree XLamp RGB LEDs, Grandar LEDArtista lighting control system. **Elements** 4,000 x ETFE 'soap bubbles' forming walls and roof. **Applications** Dramatic urban-architectural lighting displays for major events. **Appearance** Giant square building with back-illuminated, translucent, 'soap bubble' façades. **Completed** 2008 (Beijing Olympic Games). **Developer** People's Government of Beijing Municipality and Beijing State-Owned Assets Management. **Architects** CSCEC + PTW (concept Chris Bosse). **Engineers** CCDI + Arup. **Lighting Contractor** Grandar Landscape Lighting and Technology Group. **Image** Tim Griffith. **Online** water-cube.com, ptw.com.au, arup.com

Khan Shatyr Entertainment Centre, Astana, Kazakhstan p.020

Technology (as specified) Philips 250W spotlights, ERCO 35W Cylinder Façade luminaires, ERCO Visor III 35W floor washers, ERCO Focal Flood II floods. **Elements** Tripod mast and cable net-supported ETFE tent roof. **Applications** Public icon and night beacon for a desert city. **Appearance** Giant, luminous, colour-changing tent. **Completed** 2010. **Government Sponsor** President Nursultan Nazarbayev. **Developer** Sembol Construction. **Architects** Foster + Partners (Filo Russo, Peter Ridley), Linea Tusavul Architecture, Gultekin Architecture. **Structural, Mechanical and Electrical Engineers** Buro Happold (Mike Cook), with OZUN PROJE + Arce. **Lighting Consultant** Claude Engle. **Steelwork** Samko Engineering. **Cable Net** Montage Services. **ETFE Cladding** Vector-Foiltec. **Image** Nigel Young/Foster + Partners. **Online** fosterandpartners.com, crengle.com, burohappold.com

Mer-veille (at MuCEM J4), Marseille, France pp.024–025

Completed 2013. **Developer** French Ministry of Culture (DPPIC). **Venue** J4 Building, Museum of European and Mediterranean Civilizations (MuCEM). **Architects** Rudy Ricciotti (with Roland Carta). **Light Artist** AIK (Yann Kersalé). **Technical Light Art Designers** 8'18''. **Lighting Installation** SPIE. **Permanent Exhibition Lighting Design** Licht Kunst Licht. **Images** Matthieu Colin. **Online** ykersale.com, rudyricciotti.com

Yas Island Marina Hotel, Abu Dhabi, UAE pp.026–027

Completed 2009. **Developer** Aldar Properties PJSC (current operator Viceroy). **Architect** Asymptote Architecture (Hani Rashid, Lise Anne Couture). **Project Architects** Dewan Architects & Engineers, Tilke & Partners. **Structural Engineers** Dewan Architects & Engineers, Arup. **Gridshell Engineers** Schlaich Bergermann und Partner, Waagner-Biro. **MEP Engineer** Red Engineering. **Façade Consultants** Front Inc, Taw & Partner. **Gridshell BIM Consultant** Gehry Technologies. **Link Bridge Engineers** Arup Bridge, Centraal Staal. **Gridshell and Public Lighting** Arup Lighting (Brian Stacy). **Landscaping** Cracknell Landscape Architects. **Interiors** Jestico and Whiles. **Lighting Control** e:cue. **Gridshell LED Lighting** Enfis, Cooper Lighting and Safety. **Images** Arup, Bjorn Moerman, Paolo Castellani. **Online** arup.com, asymptote.net

Melbourne Theatre Company, Australia pp.028–029

Completed 2008. **Developer** Major Projects Victoria, University of Melbourne (Melbourne Theatre Company). **Lighting Design** Electrolight. **Architects** ARM Architecture. **Image** Shannon McGrath. **Online** electrolight.com.au, a-r-m.com.au

Wintergarden, Brisbane, Australia pp.030–031

Completed 2011. **Developer** ISPT. **Builder/Project Manager** Brookfield Multiplex. **Architect** Studio 505. **Lighting Contractor** Xenian. **Specialist Engineer** Tensys. **Artwork Construction** Urban Art Projects. **Video Content** Ramus Illumination. **Consulting Artist** John Warwicker. **Images** Emma Valente. **Online** studio505.com, ramus.com, xenian.com.

Silo 468, Helsinki, Finland pp.032–033

Completed 2011. **Developers** City of Helsinki Planning Department, TASKE, Helsinki Energy. **Project Manager** HKR Executive. **Designers** Lighting Design Collective. **Architect** Pöyry Finland Oy. **Electrical Engineer** Olof Granlund Oy. **Contractor** VRJ Etelä-Suomi Oy. **Images** Hannu Iso-oja, Tuomas Uusheimo, Tapio Rosenius. **Online** ldcol.com

Hatoya-3 Building, Tokyo, Japan pp.034–035

Completed 2010. **Developer** Hatoya. **Lighting Design** Forlights (Yutaka Inaba). **Architect** N˜osu Architects Planners Engineers (Yoshimitsu N˜osu). **Structural Engineering** Van Structural Design. **Electrical and Mechanical Engineers** Nichiei Architects. **Construction** Shimizu Corporation. **Images** Ryota Atarashi. **Online** nosu.jp

Fetch, Columbus, USA p.036

Completed 2010. **Venue** Wexner Center for the Arts. **Curator** Christopher Bedford. **Light Artist** Erwin Redl. **Images** Erwin Redl. **Online** paramedia.net

State Theatre Centre of Western Australia, Perth, Australia pp.036–037

Completed 2011. **Architecture** Kerry Hill Architects. **Lighting Designer** Electrolight. **Electrical Engineer** Wood & Grieve. **Lighting Suppliers** Eagle, Euroluce, ERCO. **Images** Robert Frith. **Online** electrolight.com.au, statetheatrecentrewa.com.au, kerryhillarchitects.com

Gran Casino Costa Brava, Girona, Spain pp.038–039

Completed 2010. **Developer** Gran Casino Costa Brava. **Construction Contractor** Proinosa. **Architect** b720 Fermín Vázquez Arquitectos. **Lighting Design** artec3 Studio. **Structural Engineer** BOMA. **Services Engineer** JG Asociados. **Landscape Architect** Arquitectura Agronomia. **Media Control** LEDsControl. **Lighting Suppliers** Troll. **Images** Adriá Goula, Wenzel, courtesy artec3, b720. **Online** artec3.com, b720.com

MetaLicht, Wuppertal, Germany pp.040–041

Completed 2012. **Sponsors** Bergische Universität Wuppertal, Jackstädt Stiftung, Schmersal, Vorwerk, WSW, Stadtsparkasse Wuppertal, BOROS. **Light Artist** Mischa Kuball. **Lighting Supplier** Zumtobel. **Installation** Elektro Decker. **Image** Sebastian Jarych. **Online** mischakuball.com

Blackburn Town Hall, UK p.042

Completed 2009. **Lighting Design** Studio Fink (Peter Fink). **Producer** Urban Projects. **Image** Studio Fink. **Online** studiofink.eu

Donaustrom, Vienna, Austria p.042

Completed 2007, 2008, 2010. **Host/Venue** Austrian Parliament building. **Light Artist** Waltraut Cooper. **Lighting Supplier** KLW. **Image** Robert Zolles, Austrian Parliament. **Online** cooper.at

Speicherstadt, Hamburg, Germany p.042

Completed 2001 (launch), development continuing. **Light Artist** Michael Batz. **Project Society** Licht-Kunst-Speicherstadt. **Image** Michael Batz. **Online** michaelbatz.de

Reichstag, Berlin, Germany p.043

Completed 2009. **Light Artist** Michael Batz. **Lighting Supplier** Philips. **Support** Foundation Lebendige Stadt. **Image** Michael Batz. **Online** michaelbatz.de

Capital Gate, Abu Dhabi, UAE p.044

Completed 2011. **Developer** Abu Dhabi National Exhibitions Company. **Architects, Mechanical and Electrical Consultants** Robert Matthew Johnson-Marshall & Partners (RMJM). **Lighting Design** DPA Lighting Consultants. **Lighting Suppliers** Osram (Traxon Technologies). **System Integrator** Cinmar Lighting Systems. **General Contractor** Al Habtoor Leighton Group. **Image** Jeff Schofield. **Online** rmjm.com, dpalighting.com

Burj Khalifa Tower, Dubai, UAE pp.044–045

Completed 2010. **Developer** Emaar Properties PJSC. **Project Management** NORR. **Architects** Skidmore, Owings & Merrill. **Engineers** Hyder Consulting, GHD. **Construction** Samsung Engineering and Construction, Besix, Arabtec. **Lighting Design** Fisher Marantz Stone (comprehensive scheme), Speirs + Major (celebration scheme). **Lighting Suppliers** B-K Lighting, ERCO, Cooper Lighting, Simes, WE-EF, Philips Dynalite, Tectronics, Tridonic, Oasis Enterprises. **Image** Paul Marantz. **Online** som.com, fmsp.com, speirsandmajor.com, dynalite-online.com

Cleusa Gift Store, São Paulo, Brazil pp.048–049

Completed 2008. **Developer** Santa Helena Group.

Architecture/Interiors Jayme Mestieri and Carlos Azevedo. **Lighting Design** Studio ix (Guinter Parschalk). **Construction** Matec Engenharia and Souza Lima. **Lighting Suppliers** LedPoint, Domane, Lemca, Lumini, Cynthron, Osram. **Images** Guinter Parschalk. **Online** studioix.com.br

ION Orchard, Singapore pp.050-051

Completed 2009. **Developers** CapitaLand, Sung Hung Kai Properties (Orchard Turn Developments). **Architects** Benoy Associates, RSP Architects Planners & Engineers. **Mechanical, Structural and Façade Engineering** Arup, Squire Mech. **Construction Consultants** Davis Langdon & Seah Singapore. **Images** Arup. **Online** benoy.com, arup.com

LICHT, Stuttgart, Germany pp.052-053

Completed 2010. **Host** State Bank of Baden-Württemberg (Stuttgart). **Light artist** Götz Lemberg. **Programming** TLD Planungsgruppe. **Installation** Rudolf Lichtwerbung. **Image** Manuel Dahmann. **Online** goetzlemberg.de

Weiss, Stuttgart, Germany p.053

Completed 2009. **Host** Kunstmuseum Stuttgart. **Light Artist** Götz Lemberg. **Images** Manuel Dahmann. **Online** goetzlemberg.de

Linien, Seoul, Korea p.054

Completed 2012. **Producer** Gana Art. **Light Artist** Ursula Scherrer. **Image** Ursula Scherrer. **Online** ursulascherrer.com, ganaart.com

El Molino, Barcelona, Spain pp.054-055

Completed 2010. **Developer** Ociopuro. **Architect** BOPBAA (Josep Bohigas Arnau, Francesco Pla Ferrer, Iñaki Baquero Riazuelo). **Light Design** artec3 Studio (Maurici Ginés). **Project Manager** EGI. **Installation Engineers** JMA Engineers. **Structural Engineers** BOMA. **Acoustic Engineers** Querol-Colomer. **Image** Rosa Puig. **Online** artec3.com, bopbaa.com

Miroir de Mer, Sydney, Australia pp.056-057

Completed 2013. **Developers** Fraser Properties and Sekisui House. **Light Artist** Yann Kersalé. **Architects** Jean Nouvel, PTW Architects. **Engineers** Arup Lighting (Tim Carr), Watpac. **Heliostat Contractor** Kennovations. **Lighting Contractor** Xenian. **Images** courtesy Fraser Properties. **Online** ykersale.com, jeannouvel.com

YOUR TEXT HERE, Detroit, USA pp.058-059

Completed 2012 (Detroit), New York (2012), Reykjavik (2013). **Light Artist** UNSTABLE (Marcos Zotes). **Images** Marcos Zotes. **Online** unstablespace.com

Touch. Do Not Please the Work of Art, Singapore and Sydney, Australia p.059

Completed 2010 (iLight Marina Bay festival, Singapore), 2011 (Vivid festival, Sydney). **Artist** Cornelia Erdmann. **Singapore Partner Artist** Michael Lee Hong Hwee. **Images** Jürgen Brinkmann, Cornelia Erdmann. **Online** corneliaerdmann.de, michaellee.sg

Roca Barcelona Gallery, Spain pp.060-061

Completed 2009. **Developer** Roca Sanitarios. **Architect** Office Architecture in Barcelona/OAB (Carlos Ferrater, Lucia Ferrater Arquer, Borja Ferrater Arquer). **Light Design** artec3 Studio (Maurici Ginés). **Light Control Software** Visionarte (Intesis). **Images** Aleiz Bagué, courtesy Roca. **Online** artec3.com

Code (Version 2), Beijing, China pp.062-063

Completed 2009 (Australian Film Festival). **Venue** MegaBox cinema, Sanlitun Village (now Taikoo Li Sanlitun), Chaoyang District. **Artist** Laurens Tan. **Image** Laurens Tan. **Online** laurenstan.com, octomat.com

After Light, Singapore p.063

Completed 2012 (iLight Marina Bay festival). **Producer** Storybox. **Director** Robert Appierdo. **Concept** Jess Feast. **Production Manager** Mark Westerby. **Cinematographer** Marty Williams. **Audio** Andy Hummel. **Editor** Hamish Waterhouse. **AV Tech** Stu Foster, Johann Nortje. **Production Assistant** Mary Laine. **Graphic Design** Matt Gleeson. **Video Editing** Orchard Studios. **Images** Frank Pinckers. **Online** storybox.co.nz

Sprechende Wand, Backnang and Saarbrücken, Germany p.064

Completed 2009. **Sponsor** City Gallery of Backnang. **Light Artists** Daniel Hausig, with students from Hochschule der Bildenden Künste Saar. **Images** Daniel Hausig. **Online** daniel-hausig.de

The Kind Spirits, Ljubljana, Slovenia pp.064-065

Completed 2010 (Lighting Guerrilla festival). **Light Artist** Aleksandra Stratimirovič. **Images** Damjan Kocjančič. **Online** svetlobnagverila.net, strati.se

Floating Waters, Wiesbaden, Germany pp.064-065

Completed 2013. **Venue** Zircon Tower. **Light Artist** Jens Schader (Raumbasis). **Image** Jens Schader. **Online** raumbasis.de

2-22 La Vitrine culturelle, Montreal, Canada pp.066-067

Completed 2013. **Developers** La Vitrine culturelle and Société de Développement Angus. **Media Façade Concept and Production** Moment Factory. **Architecture** Ædifica and Gilles Huot. **Equipment** Piedco. **Installation** Solotech. **Luminous Pathway** (concept) Intégral Ruedi Baur and Intégral Jean Beaudoin for Quartier des Spectacles. **Image** Moment Factory. **Online** momentfactory.com, lavitrine.com

77 Million Paintings, Sydney, Australia pp.070-071

Completed 2009 (Smart Light Sydney/Vivid festival). **Developer** Sydney Opera House. **Light Concept/Art** Brian Eno/Lumen London. **Projection Contractor** The Electric Canvas. **Multimedia Consultant** Ramus Illumination. **Images** Matt Flynn. **Online** lumenlondon.com, theelectriccanvas.com.au, ramus.com.au

Uhllich(t), Mosel Valley, Germany p.072

Completed 2010. **Host/Venue** Winninger Uhlen (Kunsttage Winningen Art Biennale). **Light Artist** Ingo Bracke. **Production** studiokybra. **Images** Jürgen Brinkmann. **Online** ingobracke.de

Felsenzauber, Bavarian Alps, Germany pp.072-073

Completed 2010 (*Aus Wasser werde Licht*), 2014 (*Felsenzauber für Nachtwandler*). **Locations** Gießenbachklamm, Kiefersfelden. **Developer** Gemeinde Kiefersfelden tourism board. **Light Artist** Ingo Bracke. **Music Compositions** Mary-Anne Kyriakou. **Production** studiokybra. **Image** Jürgen Brinkmann. **Online** ingobracke.de, felsenzauber.de

Ode à La Vie, Barcelona, Spain pp.074-075

Completed 2012. **Developers** City of Montreal with City of Barcelona. **Venue/Host** Basilica Sagrada Familia (Nativity Façade). **Concept** Renaud Architecture d'événements. **Multimedia Content/Production** Moment Factory. **Executive Producer** Éric Fournier. **Producer** Johanna Marsal. **Multimedia Director** Nelson de Robles. **Scriptwriter** Mareike Lenhart. **Creative Director** Sakchin Bessette. **Technology Director** Dominic Audet. **Creative Collaborators** Dominic Champagne, Brigitte Poupart. **Original Music** Anthony Rozankovic, Misteur Valaire. **Lighting Consultant** Dominic Lemieux. **Image** Pep Daudé/Basilica Sagrada Familia. **Online** momentfactory.com

Gutenberg, Pittsburgh, USA p.075

Completed 2008 (Festival of Lights, Pittsburgh). **Producer** ArtLumiere. **Venue/Host** Cathedral of Learning, University of Pittsburgh. **Artist** Casa Magica (Friedrich Förster, Sabine Weissinger). **Image** Casa Magica. **Online** casamagica.de

Light, Jerusalem, Israel p.076

Completed 2013 (International Festival of Light). **Curator** Eduardo Huebscher. **Venue** Old City Wall at Gan Hatekuma, Jerusalem. **Artists** Detlef Hartung and Georg Trenz. **Image** Hartung Trenz. **Online** hartung-trenz.de

Schauspiel, Fürstenfeld, Germany p.076

Completed 2011 (Brucker Kulturnacht zur Waldbühne). **Venue** Kloster Fürstenfeld, Neue Bühne Bruck. **Artists** Detlef Hartung and Georg Trenz. **Image** Hartung Trenz. **Online** hartung-trenz.de

Wortschatz – Lebenszeichen, St Goar, Germany pp.076-077

Completed 2009 (rheinpartie festival). **Curator** Helmut Bien. **Artists** Detlef Hartung and Georg Trenz. **Image** Hartung Trenz. **Online** hartung-trenz.de

Lebenszeichen, Koblenz, Germany pp.076-077

Completed 2012 (Lichtströme festival). **Curators** Bettina Pelz, Tom Groll. **Venue** Kaiser Wilhelm I Monument, Deutches Eck. **Artists** Detlef Hartung and Georg Trenz. **Image** Hartung Trenz. **Online** hartung-trenz.de

Main Embankment Panorama, Frankfurt, Germany pp.078-079

Completed 2008. **Venue/Host** European Central Bank (Frankfurt). **Light Artists** Casa Magica (Friedrich Förster, Sabine Weissinger). **Images** Casa Magica. **Online** casamagica.de

Night Beacon, Frankston, Australia p.080

Completed 2012 (ex URBAN screens festival). **Developers** City of Frankston and Frankston Arts Centre. **Light Artist** Ian de Gruchy. **Image** Nick Azidis. **Online** artprojection.com.au

Dr Who, Sydney, Australia p.080

Completed 2013 (Vivid festival). **Developers** BBC, Destination NSW. **Host/Venue** City of Sydney Customs House. **Projection Artists** Spinifex Group. **Technical Producers** Technical Direction Company (TDC). **Image** Peter Murphy. **Online** tdc.com.au, spinifexgroup.com.au

City Life, Sydney, Australia p.080

Completed 2012 (Vivid festival; curator Ignatius Jones). **Developer** Destination NSW. **Host/Venue** City of Sydney Customs House. **Projection Artists and Producers** The Electric Canvas. **Image** The Electric Canvas. **Online** theelectriccanvas.com.au

The Nights Before Christmas, Melbourne, Australia p.081

Completed 2012 (City of Melbourne Christmas projections). **Venue** Melbourne Town Hall. **Developer** City of Melbourne. **Design and Technical Production** The Electric Canvas. **Image** The Electric Canvas. **Online** theelectriccanvas.com.au

Awakening of Mary, Amsterdam, The Netherlands, and Brussels, Belgium pp.082-083

Completed 2012 (Old Church, Amsterdam), 2013 (Notre Dame de la Chapelle, Brussels). **Light Artist** Titia Ex. **Music** Nederlands Kamerkoor. **Image** Titia Ex. **Online** titiaex.nl

CCTV/Creative Control, New York, USA p.083

Completed 2011. **Artist** UNSTABLE (Marcos Zotes). **Image** Marcos Zotes. **Online** unstablespace.com

Rafmögnuð Náttúra, Reykjavik, Iceland p.083

Completed 2012 (Winter Lights festival). **Creative Direction** UNSTABLE (Marcos Zotes). **Visual Production** Marcos Zotes, Andrea Dart, Chris Jordan, Thessia Machado, Steven Tsai, Noa Younse. **Choreography and Dance Performance** Coco Karol. **Music** For a Minor Reflection. **Project Management** Marcos Zotes, Gerdur Sveinsdóttir. **Technical Direction** Chris Jordan. **Technical Production** Luxor. **Videography** Azmi Mert Erdem. **Images** Jón Óskar Hauksson, Hallgrímur Helgasson. **Online** unstablespace.com, rafmognudnattura.com

Sky Machine, Toruń, Poland p.084

Completed 2009. **Venue** Holy Spirit Church, Toruń. **Sponsor** Torunska Agenca. **Light Artist** Ocubo (Nuno Maya, Carole Purnelle). **Music** Luis Cília. **Image** Ocubo. **Online** ocubo.com

Sintra Garrett, Portugal pp.084-085

Completed 2011 (Lumina festival). **Venue** National Palace, Sintra. **Sponsor** CMSintra. **Light Artist** Ocubo (Nuno Maya, Carole Purnelle). **Music** Luis Cília. **Image** Ocubo. **Online** ocubo.com

Project Credits

The City of My Dreams, Amsterdam, The Netherlands pp.084-085

Completed 2010 (Amsterdam Light Festival). **Venue** Hermitage for Children. **Light Artist** Ocubo (Nuno Maya, Carole Purnelle). **Music** Sylvain Moreau. **Image** Ocubo. **Online** ocubo.com

Projecting April, Lisbon, Portugal pp.084-085

Completed 2014 (celebrations for the 30th anniversary of the Revolution of Carnations). **Venue** Terreiro do Paço. **Sponsor** EGEAC. **Light Artist** Ocubo (Nuno Maya, Carole Purnelle). **Music** Luis Cília. **Image** Ocubo. **Online** ocubo.com

Time Drifts, Montreal, Canada pp.086-087

Completed 2010 (Mutek festival). **Location** Place des Festivals. **Artist** Philipp Geist Studio. **Image** Philipp Geist. **Online** p-geist.de

Time Drifts, Frankfurt, Germany p.087

Completed 2012 (Luminale festival; curator Helmut Bien). **Venue** Kulturcampus, Goethe University. **Artist** Philipp Geist Studio. **Image** Philipp Geist. **Online** p-geist.de

Lighting Up Times, Barfüßerkirche Erfurt, Germany p.087

Completed 2012 (Erfurt 2012 festival). **Artist** Philipp Geist Studio. **Image** Philipp Geist. **Online** p-geist.de

Construction-Deconstruction, Otterndorf, Germany p.087

Completed 2011. **Venue** Museum Gegenstandsfreier Kunst. **Artist** Philipp Geist Studio. **Image** Philipp Geist. **Online** p-geist.de

Bow West Olympics VIP Entry, London, UK pp.088-089

Completed 2012. **Developer** Olympic Delivery Authority. **Lighting Designer** Speirs + Major (Philip Rose, Ting Ji). **Architects** Allies and Morrison. **Main Contractor** Bankside. **Lighting Equipment Supplier** Enliten. **Image** Ting Ji/Speirs + Major. **Online** speirsandmajor.com

Visual Piano, Lüdenscheid, Germany pp.090-091

Completed 2013 (LightRouten festival). **Developer** Lüdenscheider Stadtmarketing. **Curators** Bettina Pelz, Tom Groll. **Artists** Theinert-Scherrer (Kurt Laurenz Theinert on piano, Ursula Scherrer on visuals). **Visual Piano Software** Roland Blach, Philip Rahlenbeck. **Other collaborations** *Hammerhaus* (Theinert with Hanfreich), *Klangvisionen* (Theinert with Tuna Pase), *Die Sonografen* (Theinert with Fried Dähn). **Images** Theinert-Scherrer. **Online** theinert-lichtkunst.de

Columbus 2.0, Seville, Spain pp.092-093

Completed 2008 (Sevilla Bienal), 2011 (Thessaloniki, Greece, and Karlsruhe, Germany). **Light Artists** Michael Bielicky, Kamilla B. Richter. **Project team** Andrej Jungnickel, Dirk Reinbold, Andreas Siefert. **Image** courtesy Michael Bielicky. **Online** bielicky.net

Falling Times, various locations p.093

Completed 2007 (YOU-ser, ZKM Center, Karlsruhe, Germany), 2008 (Mostra SESC des Artes, São Paulo), 2009 (Bohemian National Hall Czech Center, New York). **Light Artists** Michael Bielicky, Kamilla B. Richter, Dirk Reinbold. **Image** courtesy Michael Bielicky. **Online** bielicky.net

Organic TV, Brighton, UK p.093

Completed 2013 (Brighton Digital Festival). **Host/Venue** Phoenix Gallery. **Curators** Sue Gollifer, Karin Mori. **Artist** William Latham. **Projectionist** Nick Fenwick. **Software** Stephen Todd. **Image** Nick Fenwick. **Online** latham-mutator.com, phoenixbrighton.org

The Irreversible, Ljubljana, Slovenia pp.094-095

Completed 2010 (solo exhibition at Kapelica Gallery; curator Jurij Krpan). **Artist** Norimichi Hirakawa. **Image** courtesy Norimichi Hirakawa. **Online** counteractiv.com

Celebration of Life, Singapore p.108

Technology 2 x Coolux Pandoras Box QUAD media servers (one back-up), 16 x Panasonic PT-DZ21K projectors, Lightware fibre optic cables. **Completed** 2014 (iLight Marina Bay festival). **Developer** Singapore Urban Redevelopment Authority. **Producer** Pico Art International. **Curators** Ong Swee Hong, Andrew Lee, Tai Lee Siang. **Light Artist** Justin Lee. **Technical Design** Dorier Asia. **Collaborators** Marina Bay Sands, Panasonic Systems. **Image** courtesy URA. **Online** ilightmarinabay.sg, justinleeck.com, dorier-group.com, coolux.de

Passing Through Light, Charlotte, USA pp.112-113

Completed 2012. **Curator** Jean Greer. **Artist** Erwin Redl. **Image** Erwin Redl. **Online** paramedia.net

Voyage, London, UK, and Scottsdale, USA pp.114-115

Completed 2012 (London) and 2013 (Scottsdale, Arizona). **Light Artists** Aether & Hemera. **Images** Philip Vile, Peter Matthews, Ian Docwra. **Online** aether-hemera.com

Fancy/Lightweight, Singapore pp.116-117

Completed 2012. **Light Artist** Cornelia Erdmann. **Images** Cornelia Erdmann. **Online** corneliaerdmann.de

Bibigloo, various locations p.118

Completed 2010 (Wirksworth Festival, UK), 2012 (iLight Marina Bay festival, Singapore, and Vivid festival, Sydney, Australia), 2013 (Lumina festival, Cascais, Portugal) and 2014 (Circle of Light festival, Moscow, Russia). **Artist** BIBI (Fabrice Cahoreau). **Production** Agence Tagada. **Images** courtesy BIBI. **Online** bibi.fr

Le Roi des Dragons, Lyon, France, and Dubai, UAE pp.118-119

Completed 2012 (Fête des lumières, Lyon), also shown 2014 (Festival of Lights, Dubai). **Artist** BIBI (Fabrice Cahoreau). **Production** Agence Tagada. **Image** courtesy BIBI. **Online** bibi.fr

The Waiting, Vlieland, The Netherlands pp.120-121

Completed 2010 (Into the Great Wide Open festival;

curator Carolien Euser). **Light Artist** Titia Ex. **Image** Titia Ex. **Online** titiaex.nl

Fairy Lake Kayak, Huntsville, Canada p.121

Completed 2014. **Location** Fairy Lake, Huntsville, Ontario. **Photographic Artist** Stephen Orlando. **Collaborator** Jessica Bruce. **Image** Stephen Orlando. **Online** motionexposure.com

Chimney Corner #2, Cape Breton Island, Canada p.121

Completed 2010. **Light and Photographic Artist** Vicki DaSilva. **Image** Vicki DaSilva. **Online** vickidasilva.com

Urban Green, Stockholm, Sweden p.122

Completed 2011. **Developer** Stockholm City. **Concept and Light Artist** Ljusarkitektur, now ÅF Lighting (Isabel Villar, Joonas Saaranen). **Sound** Tyréns (Clas Torehammar), Urban Sound Institute (Björn Hellström), Sennheiser Nordic. **Construction** Elfströms El. **Sponsors** Santa & Cole (fixtures and furniture), Dahl Agenturer. **Image** Joonas Saaranen. **Online** af-lighting.com

rocklights, Sydney, Australia p.122

Completed 2009 (Smart Light Sydney/Vivid festival). **Venue** Argyle Cut, The Rocks. **Light Artist** Ingo Bracke. **Image** Mat Flynn. **Online** ingobracke.de

Human Effect, Melbourne and Sydney, Australia pp.122-123

Completed 2012 (Melbourne Festival) and 2013 (Vivid festival, Sydney). **Locations** Lingham Lane (Melbourne), Nurses Walk (Sydney). **Artist** Yandell Walton. **Animation** Tobias Edward. **Software** Jayson Haebich. **Images** Lauren Dunn. **Online** yandellwalton.com.

The Nyborg Bridges, Funen, Denmark pp.124-125

Completed 2010. **Developer** The Danish Road Directorate (Niels Juels Gade). **Lighting Designers** ÅF Lighting (Frida Nordmark, Frederik Waneck Borello, Christian Klinge, Allan Ruberg). **Images** Lars Bahl. **Online** af-lighting.com

Aspire, Sydney, Australia p.125

Completed 2010. **Developer** City of Sydney. **Artist** Warren Langley (project manager Trent Baker). **Image** Richard Glover. **Online** warrenlangley.com.au

Q150, Brisbane, Australia pp.126-127

Developer Brisbane City Council. **Curator** Museum of Brisbane (Louise Rollman). **Concept Artist** Ian de Gruchy. **Images** David Sandison, Morris Weyer. **Online** artprojection.com.au, kateshaw.org, laithmcgregor.blogspot.com

Glühwürmchen Project, various locations pp.128-129

Completed 2009 (*Quantum Flowers*, Ludwigsburg, Germany), 2010 (*The Fireflies Factory*, VSL Lindabrunn, Vienna), and 2010 (*The Fireflies Fence*, Pedrinate frontier between Italy and Switzerland; iLight Marina Bay festival, Singapore). **Light Artist** Francesco

Mariotti. **Images** courtesy Francesco Mariotti. **Online** mariotti.ch, gluehwuermchen.ch

Crystallized, Sydney, Australia, and Singapore p.129

Completed 2011 (Vivid festival, Sydney) and 2012 (iLight Marina Bay festival, Singapore) **Light Artists** Andrew Daly, Katharine Fife. **Images** Daly and Fife. **Online** andrewdaly.com, katharinefife.com

Temporary Temple Pavilion, Hooghly, India pp.132-133

Completed 2012. **Developer** Kishor Sangha Community. **Architects/Light Artists** Abin Design Studio (Abin Chaudhuri and Tilak Ajmera). **Images** Abin Chaudhuri. **Online** abindesignstudio.com

Pulse Park, New York, USA pp.134-135

Completed 2008 (also Bochum, 2012). **Developer** Madison Square Park Conservancy (with funds from New York City Department of Cultural Affairs). **Light Artist** Rafael Lozano-Hemmer (Antimodular). **Programming** Conroy Badger. **Staging** Scharff Weisberg. **Antimodular Production Team** David Lemieux, Natalie Bouchard, Boris Dempsey, Stephan Schulz, Pierre Fournier. **Madison Square Park Production Team** Debbie Landau, Sam Rauch, Jeffrey Sandgrund, Stewart Desmond. **Images** James Ewing. **Online** lozano-hemmer.com

Helsingborg Waterfront, Sweden pp.136-137

Completed 2010. **Developer** Helsingborg Municipality. **Lighting Designer** ÅF Lighting (F. Hansen & Henneberg). **Landscape Design** Brandt Landscape. **Lighting Suppliers** Philips, Aubrilam, Bico, iGuzzini, Roblon. **Image** Martin Nordmark (F. Martin Kristiansen). **Online** af-lighting.com

Aalborg Harbour, Denmark pp.136-137

Completed 2011. **Developer** Aalborg Municipality. **Lighting Designer** ÅF Lighting (F. Hansen & Henneberg). **Architects** CF Möller. **Environmental Planning** COWI A/S. **Landscape Design** Vibeke Rønnow. **Lighting Suppliers** Sill, Philips, iGuzzini, Simes, Roblon. **Images** Martin Nordmark (F. Martin Kristiansen), Ole Mikael Sørensen. **Online** af-lighting.com

Ishøj Station Plaza, Copenhagen, Denmark p.137

Completed 2012. **Developer** Ishøj Municipality. **Architects** Arkitema Architects. **Lighting Designer** ÅF Lighting (F. Hansen & Henneberg). **Contractor** Zacho-Lind. **Electrical Contractor** Bravida Danmark. **Specialist Consultants** VIA Trafik, Atkins Dines Jørgensen, Ginsberg Madsen. **Lighting Suppliers** Louis Poulsen, Philips, ERCO, iGuzzini, BEGA. **Image** Martin Kristiansen. **Online** af-lighting.com

El Circuito Mágico del Agua, Lima, Peru pp.138-139

Completed 2007 (updating original water works completed by engineer Alberto de Jaxas Jochamowitz and architect Claudio Sahut for the government of President Augusto B. Leguía in 1929). **Location** Parque de la Reserva. **Developer/Operator** Empresa Municipal Inmobiliaria de Lima (EMILIMA). **Images** Parque de la Reserva. **Online** parquedelareserva.com, emilima.com.pe

Mingus Streetlamps, Le Havre, France pp.140-141

Completed 2012. Developer City of Le Havre. Project Supervisor Ateliers Lion Associés. Streetlamp Design Atelier H. Audibert. Fabrication Inédit Lighting for Santa & Cole. Image Giacomo Bretzel. Online atelierherveaudibert.com

Firalet Square, Olot, Spain pp.140-141

Completed 2011. Developer Ajuntament d'Olot. Architects RCR Arquitectes. Lighting Design artec3 Studio. Image Pep Sau. Online artec3.com, rcrarquitectes.es

The High Line, New York, USA pp.142-143

Completed 2011-14. Developer New York City Economic Development Corporation. Architects Diller Scofidio + Renfro. Landscape Architecture James Corner Field Operations. Lighting Design L'Observatoire International (Hervé Descottes). Structural and MEP Engineering Buro Happold. Environmental Engineering GNB Services. Structural Engineering and Historic Preservation Robert Silman Associates. Public Space Management ETM Associates. Planting Design Piet Oudolf. Images Iwan Baan. Online thehighline.org

Reflecting Absence, New York, USA pp.144-145

Completed 2011. Developer Lower Manhattan Development Corporation. Owner National September 11 Memorial and Museum. Construction Manager Port Authority of New York and New Jersey. Lighting Design Fisher Marantz Stone (Paul Marantz, Zack Zanolli, Carla Ross-Allen, Barry Citrin). Architect Michael Arad. Landscape Architecture Peter Walker & Partners. Images Caridad Sola, Port Authority of New York and New Jersey, Fisher Marantz Stone. Online fmsp.com

1.26, Singapore and other locations pp.146-147

Completed 2014 (iLight Marina Bay festival; curators Ong Swee Hong, Andrew Lee, Tai Lee Siang); also 2010 ('Biennial of the Americas' exhibition, Denver Art Museum), 2011 (Art&About festival, Sydney), 2012-13 (Amsterdam Light Festival). Sponsor Singapore Urban Redevelopment Authority. Artist Janet Echelman. Design Engineer Peter Heppel & Associates. Fibre (Spectra) Supplier Honeywell. Original Project Sponsor Denver Office of Cultural Affairs. Singapore Lighting Design Martin Professional, Kurihara Kogiyo. Image courtesy Singapore Urban Redevelopment Authority. Online echelman.com

Sunken Garden, London, UK, and other locations p.147

Completed 2014. Venue Kensington Palace Gardens. Light Artist Paul Friedlander. Images courtesy Paul Friedlander. Online paulfriedlander.com

Into the Blu, Ljubljana, Slovenia pp.148-149

Completed 2009 (Nuit Blanche, Brussels, and Fête des lumières, Lyon), 2010 (Noche Blanca, Bilbao), 2011 (Lighting Guerrilla festival, Ljubljana, and Belgrade of Light, Belgrade). Producers Lighting Guerrilla team and Arts Council Pro Helvetia. Light Artist Sophie Guyot. Image Unknown. Online sophieguyot.ch

Lines up. a recollection, Berlin, Germany pp.148-149

Completed 2012. Light Artist Jeongmoon Choi

(represented by galerie laurent mueller, Paris). Image Jeongmoon Choi. Online jeongmoon.de

The Pool, Black Rock City, USA pp.148-149

Completed 2008 (Burning Man festival, Black Rock City), 2014 (iLight Marina Bay festival, Singapore, and AHA! Light Up Cleveland festival, USA). Sponsors Black Rock Arts Foundation. Light Art Jen Lewin Studio. Collaborators Bill Magnusson, Dan Julio Designs. Image Jürgen Brinkmann. Online jenlewinstudio.com

My Public Garden, Singapore pp.150-151

Completed 2010 (iLight Marina Bay festival, Singapore); other versions completed for other cities. Developer Singapore Urban Redevelopment Authority. Light Artists TILT (François Fouilhé and Jean-Baptiste Laude). Image courtesy Singapore Urban Redevelopment Authority. Online t-i-l-t.com

La Fontaine aux Poissons, Lyon, France pp.150-151

Completed 2008 (Fête des lumières). Artist BIBI (Fabrice Cahoreau). Production Agence Tagada. Technical Specialist Magnum. Image courtesy BIBI. Online bibi.fr

BIBI's Hell, It is Here, Geneva, Switzerland p.151

Completed 2012 (Festival des arbres et lumières, Geneva); earlier versions 2001 (59, rue de Rivoli, Paris) and 2004 (Villeneuve les Maguelones, France). Artist BIBI (Fabrice Cahoreau). Production Agence Tagada. Images courtesy BIBI. Online bibi.fr

Darling Quarter Children's Playground, Sydney, Australia p.152

Completed 2012. Owner NSW Government. Development Team Lend Lease. Landscape Architects ASPECT Studios. Lighting Design Lend Lease with Speirs + Major. Environmental Graphics Deuce Design. Images courtesy Lend Lease. Online speirsandmajor.com, aspect.com.au, lendlease.com

Luminous at Darling Quarter, Sydney, Australia pp.152-153

Completed 2012. Development Team Lend Lease, Commonwealth Bank, Sydney Harbour Foreshore Authority. Lighting Design and Art Lend Lease with Ramus Illumination. Architects fjmt. Construction Lend Lease. Property Manager JLL for Darling Quarter Joint Owners. Images Robin Thompson. Online ramus.com.au, lendlease.com

Ice House Square, Swansea, Wales p.154

Completed 2012. Developer City and County of Swansea. Design Studio Fink (Peter Fink). Lighting Installation Urban Projects. Images Studio Fink. Online studiofink.eu

BruumRuum!, Barcelona, Spain pp.154-155

Completed 2013. Developer Ajuntament de Barcelona (BIMSA). Location Plaça de les Glòries. Architects MBM Arquitectes. Artist David Torrents. Lighting Design artec3 Studio (Maurici Ginés). DMX Control System Rebeca Sánchez (Leds Control). Images Ramón Ferreira (artec3). Online artec3.com

Packaged River, Caracas, Venezuela pp.158-159

Completed 2012. Light Artists Luz Interruptus. Collaborators Bicycle Workshop, Cachao Culture, Green Banana. Images Gustavo Sanabria, courtesy Luz Interruptus. Online luzinterruptus.com

Preparations for a Possible Future, Hagen, Germany pp.160-161

Completed 2010 (Ruhr: European Capital of Culture celebrations). Developers/Sponsors Stadt Hagen, Fonds BKVB, Generalkonsulat der Niederlande. Light Artist Geert Mul. Project Team Wiljnand Veneberg, Gyz La Riviere, Tim van Cromvoirt. Images Geert Mul, Wiljnand Veneberg. Online geertmul.nl

Välkommen hem, Stockholm, Sweden pp.162-163

Completed 2013. Light Artist Aleksandra Stratimirovic̈. Assistants Carl-Johna Nordin, Jonas Bard, Måns Ejhed. Architect Nivå Landskapsarkitektur. Art Curator Birgitta Silfverhielm. Images Robin Hayes. Online strati.se

The Helix Bridge, Singapore pp.164-165

Completed 2010. Developer Singapore Urban Redevelopment Authority. Architects Cox with Architects 61. Structural, Civil, Maritime, Mechanical, Electrical Engineering and Lighting Design Arup. Specialist Consultants Sato Kogyo, TTJ. Image Darren Soh. Online arup.com, coxarchitecture.com

WattFish?, Singapore pp.164-165

Completed 2010 (iLight Marina Bay festival; curators Mary-Anne Kyriakou, Kelley Cheng, Randy Chan). Light Art Dan Foreman, Cherry Wang, Zi Chang Lee (Meinhardt Light Studio). Image Jürgen Brinkmann for Smart Light Singapore. Online meinhardtls.com

Kurilpa Bridge, Brisbane, Australia pp.164-165

Completed 2009. Developer Baulderstone. Architects Cox Rayner. Structural, Civil, Geotechnical and Lighting Engineer Arup. Image Arup. Online arup.com, coxarchitecture.com

Sheikh Zayed Bridge, Abu Dhabi, UAE p.165

Completed 2011. Developer Sheikh Sultan Bin Zayed Al Nahyan and Abu Dhabi Municipality (lighting expert Martin Valentine). Architect Zaha Hadid Architects (Graham Modlen). Structural Engineer Arup. Lighting Design Arup (Rogier van der Heide, Imke van Mil, Simone Collon), Hollands Licht Advanced Lighting Design (Bob van der Klaauw). Lighting Equipment Supplier Martin Professional Middle East, High Point Rendel, Sixco, Philips Lighting, Archirodon. Images Christian Richters, courtesy Arup. Online arup.com, zaha-hadid.com

Broken Light, Rotterdam, The Netherlands pp.166-167

Completed 2010. Developers Commune Rotterdam, DSPS, De Player, Center Arts & Pact op Zuid Rotterdam. Light Designers/Artists Daglicht & Vorm (Rudolf Teunissen, Marinus van der Voorden). Streetlight Specialists Max Designers (luminaire housing design), Modernista (light calculations), Alanod (reflector efficiency). Images Studio Hans Wilschut, Rudolf Teunissen. Online brokenlight.org

Salon, Belgrade, Serbia/Sweet Home, Singapore p.168

Completed 2011 (Belgrade of Light festival), 2012 (iLight Marina Bay festival, Singapore). Light Artist Aleksandra Stratimirovi̇c̈. Image Stanislav Milojkovi̇c̈. Online strati.se

Cumulus, Sydney, Australia p.168

Completed 2012 (Vivid festival; curator Anthony Bastic). Venue Mill Lane, The Rocks. Light Artists Ruth McDermott, Ben Baxter. Lighting Installation Xenian Lighting (Philips Color Kinetics). Image Ben Baxter. Online mcdermottbaxter.com

Hopscotch, Sydney, Australia pp.168-169

Completed 2011 (Vivid festival; curator Anthony Bastic). Light Artists Fiona Venn, Reinhard Germer (Meinhardt Light Studio). Equipment Sponsors Klik Systems, RS Components. Electrical Installation Star Group. Image Klik Systems. Online meinhardtls.com

Mersey Wave, Liverpool, UK pp.170-171

Completed 2009. Developers Speke Garston, Liverpool Land Development. Sponsor Jaguar Cars. Lighting Designer Studio Fink (Peter Fink). Image Studio Fink. Online studiofink.eu

Twin Sails Bridge, Poole, UK p.171

Completed 2012. Developer Borough of Poole. Architect Wilkinson Eyre. Lighting Designer Speirs + Major. Electrical Engineer Ramboll. Principal Contractor Hochtief (UK) Construction. Electrical Contractor IES. Image Dave Morris. Online wilkinsoneyre.com, speirsandmajor.com

LightBridge, Cambridge–Boston, USA p.171

Completed 2011 (MIT 150th anniversary celebration). Lighting and Interaction Designers Susanne Seitinger and Pol Pla. Project Team Russell Cohen, Eugene Sun, Andrew Chen, Dave Lawrence, Daniel Taub, David Xiao. Sponsors Philips Color Kinetics, MIT, UROP Office, MIT Council of the Arts, MIT Festival of Art, Science and Technology, Cisco, Panasonic, SparkFun Electronics, with more than 80 volunteers. Image Peter Schmitt. Online labcast.media.mit.edu

Bridge at Cherry Orchard Cemetery, Yilan, Taiwan pp.172-173

Completed 2010. Architects Fieldoffice Architects. Light Artist CMA Lighting Design. Images CMA Lighting. Online cmalighting.com, fieldoffice-architects.com

Appears@Amsterdam, The Netherlands p.173

Completed 2012 (Amsterdam Light Festival). Light Artist Titia Ex. Images Taco Anema. Online titiaex.nl

Current 3 (cubed), Denver, USA p.173

Completed 2009. Organizer LoDoLights (Diane Huntress with Lower Downtown Neighborhood Association, Denver, CO). Installation Artist/Light Designer Virginia Folkestad with 186 Lighting Design Group. Image Diane Huntress. Online lodolights.org, virginiafolkestad.com

Regent Street Relight, London, UK pp.174-175

Completed Ongoing since 1995. **Developer** The Crown Estate. **Lighting Design** Tony Rimmer (Studio-29, formerly Imagination). **Principal Lighting Supplier** Commercial Lighting **Maintenance Contractor** IDWe. **Principal Luminaire Manufacturer** Willy Meyer + Sohn. **Images** James Newton, Tony Rimmer, Andy Spain. **Online** studio-29.co.uk

Parmenides I (Star Geode), Los Angeles, USA p.184

Technology 4 x BenQ SP870 5000 lumens projectors, media server with nVidia GPU, custom software built with Derivative Touch Designer. **Elements** Aluminium fabrication. **Applications** Interactive sculpture with projected imagery. **Appearance** Optical projections on a polyhedron. **Completed** 2011 ('Astral Flight Hangar' exhibition at Christopher Henry Gallery, Los Angeles), 2012 (iLight Marina Bay festival, Singapore; curator Mary-Anne Kyriakou). **Light Artist** Dev Harlan. **Image** Jürgen Brinkmann for Smart Light Singapore. **Online** devharlan.com

Equación Solar, Melbourne, Australia pp.188-189

Completed 2010. **Host** Light in Winter festival (Robyn Archer). **Venue** Federation Square (Kate Brennan). **Light Artist** Rafael Lozano-Hemmer. **Programming** Conroy Badger. **Antimodular Production** David Lemieux, Karine Charbonneau, Stephen Schulz, Guillaume Tremblay, Natalie Bouchard, Gideon May, Susie Ramsay. **Images** courtesy Rafael Lozano-Hemmer. **Online** lozano-hemmer.com

The Walk, Eindhoven, The Netherlands, and London, UK p.189

Completed 2012 (GLOW festival, Light Art Centre, Eindhoven), 2013 (Kinetica Museum, London). **Artist** Titia Ex. **Lighting Supplier** Philips Color Kinetics. **Image** Titia Ex. **Online** titiaex.nl

Keyframes, Lyon, France pp.190-191

Completed 2011 (Fête des lumières, Lyon, and other cities/venues later). **Light Artists** Groupe LAPS. **Concept Director** Thomas Veyssière. **Pipeline Concept** Pierre Froment. **Sound Designer** Erwan Quintin. **Senior Keyframer/3D Artist** Richard le Bihan. **Junior Keyframer/3D Artist** Vincent Sevoz. **Software Designer** (WhiteCAT) Christophe Guillermet. **Construction** Olivier Fermier, Emmanuel Céalis. **Images** Groupe LAPS. **Online** groupe-laps.org

Digital Origami Tigers, Sydney, Australia, and other locations pp.192-193

Completed 2010 (Chinese New Year celebrations, Sydney). **Producer** City of Sydney Customs House (curator Jennifer Kwok). **Architects** Laboratory for Visionary Architecture (LAVA). **Lighting Supplier** Xenian (Philips Color Kinetics). **Textile Supplier** Barrisol. **Fabrication** Sydney Stretch Ceilings. **Images** Peter Murphy (courtesy City of Sydney Customs House), Patrick Bingham-Hall. **Online** l-a-v-a.net

The Golden Moon, Hong Kong p.193

Completed 2012 (Lee Kum Kee Lantern Wonderland celebration). **Light and Sound Design** LEDARTIST (Teddy Lo). **Architectural Design** L.E.A.D. **Images** LEDARTIST. **Online** ledartist.com

Positive Attracts, Singapore, and Sydney, Australia p.194

Completed 2010 (iLight Marina Bay festival, Singapore) and 2011 (Vivid festival, Sydney). **Light Artist** Edwin Cheong. **Image** Jürgen Brinkmann for Smart Light Singapore. **Online** edwincheong.com

Guardians of Time, Berlin, Germany, and other locations pp.194-195

Completed 2011 (Festival of Lights, Berlin; first installation 2006 at Sculpture Artpark, Linz, Austria; various other appearances). **Light Artist** Manfred Kielnhofer. **Image** courtesy Manfred Kielnhofer. **Online** kielnhofer.at

A Blue Mirage in the City of Light, Singapore pp.196-197

Completed 2010 (iLight Marina Bay festival; curators Mary-Anne Kyriakou, Kelley Cheng, Randy Chan). **Developer** Singapore Urban Redevelopment Authority. **Artists** WY-TO Architects (Yann Follain, Pauline Gaudry). **Structure** AE Models. **Logistics and Electrical Agility. Image** Jürgen Brinkmann for Smart Light Singapore. **Online** wy-to.com

Dar Luz, Eindhoven, The Netherlands p.197

Completed 2008 (GLOW festival; curators Bettina Pelz, Tom Groll). **Artists** Ali Heshmati (Laboratory for Environments, Art and Design), Lars Meeß-Olsohn. **Sound** Andreas Pasika. **Technical Support** Matthias Boeser. **Images** Ali Heshmati, Claus Langer. **Online** leadinc.no

Mobile Orchard, London, UK pp.198-199

Completed 2013 (City of London festival; curator Emma McGovern). **Venue Hosts** Broadgate Estates, Devonshire Square Management, Land Securities, 30 St Mary Axe Management. **Sponsors** Bloomberg, DHH Timber, Tellings Transport, YouGarden, Worshipful Company of Fruiterers. **Architect-Artist** Atmos (Alex Haw). **Structural Engineering** Blue Engineering (James Nevin). **Lighting Design** Arup Lighting (Arfon Davies). **Fabrication** Nicholas Alexander. **Lighting Suppliers** LED Linear, Wibre, Architectural FX, Stuart Knox. **Images** Alex Haw, Jonathan Perugia. **Online** atmosstudio.com, mobileorchard.info

Illumination Disorders II, Singapore p.200

Completed 2012 (iLight Marina Bay festival; curators Mary-Anne Kyriakou, Charmaine Toh, FARM). **Artist** Tay Swee Siong. **Electronics Supplier** Adafruit. **Image** Tay Swee Siong. **Online** sweesiong.blogspot.co.uk

Orkhˉestra, Frankfurt, Germany, and Sydney, Australia pp.200-201

Completed 2014 (Luminale festival, Frankfurt, curator Helmut Bien; and Sydney Design Festival). **Lighting Designer** M. Hank Haeusler with Media Architecture Institute (MAI), Computational Design (CoDe) at University of New South Wales, Städelschule Architecture Class (SAC), Ludwig Maximilian University. **Electronics Supplier** AHL LEDs. **Images** Wolfgang Leeb. **Online** mediarchitecture.org

Lotus Dome, Lille, France, and other locations pp.202-203

Completed 2011 ('Fantastic' exhibition at Lille3000 promotion; curator Caroline David). **Original Developer** City of Lille. **Original Host/Venue** Sainte Marie Madeleine Church. **Light Artist** Studio Roosegaarde (Daan Roosegaarde). **Images** Studio Roosegaarde. **Online** studioroosegaarde.net

C/C, Singapore, Sydney, Australia, and Amsterdam, The Netherlands p.204

Completed 2010 (iLight Marina Bay festival, Singapore; curators Mary-Anne Kyriakou, Kelley Cheng, Randy Chan; also shown 2011 (Vivid festival, Sydney; curator Anthony Bastic) and 2012 (Amsterdam Light Festival). **Developer** Singapore Urban Redevelopment Authority. **Light Artist** Angela Chong. **Image** Jürgen Brinkmann. **Online** angelachong.wix.com

Flower from the Universe, various locations pp.204-205

Completed 2010 (Blob festival, Eindhoven), 2010 (Natural History Museum Senckenberg, Frankfurt), 2011 (Glass and Light Biennale, Kijkduin, The Hague), 2012 (Hortus Botanicus, Amsterdam), 2012 (Kinetica, London), 2013 (Light in Jerusalem festival). **Light Artist** Titia Ex. **Images** Peter Cox, Oren Cohen, Titia Ex, Luuk Kramer. **Online** titiaex.nl

e|MERGence, Sydney, Australia p.206

Completed 2014 (Vivid festival; curator Anthony Bastic). **Light Artists, Sculpture, Programming, Cabinetry** The Buchan Group (Anthony Rawson, Patrick Shirley, Gary Edmonds, Daniel Thomas, Lincoln Savage). **CNC Routing** John Cox's Creature Workshop. **Image** courtesy Destination NSW. **Online** buchan.com.au

Light of the Merlion, Singapore pp.206-207

Completed 2012 (iLight Marina Bay festival; curators Mary-Anne Kyriakou, Charmaine Toh, FARM). **Light Artists** Ocubo (Nuno Maya, Carole Purnelle). **Images** Ocubo. **Online** ocubo.com

Techno Nature: Bacillus, Hong Kong pp.208-209

Completed 2013 ('Imminent Domain: Designing the Life of Tomorrow' exhibition). **Event Organizer** Asia Society. **Light Artist** LEDARTIST (Teddy Lo). **Image** LEDARTIST. **Online** teddylo.tv

Cloud, Calgary, Canada, and other locations pp.208-209

Completed 2012 (Nuit Blanche festival, Calgary), 2013 (Art Experiment, Moscow), 2014 (iLight Marina Bay festival, Singapore), and other locations. **Light Artists** Caitlind R.C. Brown, Wayne Garrett. **Image** Singapore Urban Redevelopment Authority. **Online** incandescentcloud.com

Jellight, Sydney, Australia, and Singapore pp.208-209

Completed 2009 (Vivid festival, Sydney; curator Mary-Anne Kyriacou), 2010 (iLight Marina Bay festival, Singapore). **Light Artists** Pascal Petitjean, Aamer Taher, Simon Lee. **Helium Balloons** Airstar-Asia. **Image** Jürgen Brinkmann/Smart Light Sydney. **Online** airstar-asia.com, aamertaher.com

Andrea, Richmond, USA p.209

Completed 2012 ('Inlight Richmond' exhibition at 1708

Gallery). Light Artist Jason Peters. **Image** Jason Peters. **Online** jasonpeters.com

The Beginning, Singapore p.209

Completed 2009 (Singapore Night Festival). **Venue** National Museum of Singapore. **Light Artist** Sun Yu-Li. **Image** Sun Yu-Li. **Online** sunyuli.com

Urban Light, Los Angeles, USA pp.210-211

Completed 2008. **Host/Venue** Los Angeles County Museum of Art (LACMA). **Artist** Chris Burden. **Image** Chris Burden, courtesy Gagosian Gallery and Museum Associates/LACMA. **Online** lacma.org, gagosian.com

Daylight Flotsam Venice, Italy pp.210-211

Completed 2013 (New Zealand's 'Front Door Out Back' exhibition at the Venice Biennale). **Venue** Istituto Santa Maria della Pietà. **Developers** Creative New Zealand, Arts Council of NZ Toi Aotearoa. **Commissioner** Jenny Harper. **Curator** Justin Paton. **Artist** Bill Culbert. **Image** Jennifer French, courtesy Creative NZ. **Online** nzatvenice.com

Doves That Cry, Sydney, Australia, and Singapore p.211

Completed 2009 (Smart Light Sydney/Vivid festival; curator Mary-Anne Kyriakou) and 2010 (iLight Marina Bay festival; curators Mary-Anne Kyriakou, Kelley Cheng, Randy Chan). **Venues** City of Sydney Customs House and Marina Bay City Gallery. **Light Artists** Mary-Anne Kyriakou, Joe Snell. **Music Composer** Mary-Anne Kyriakou. **Images** Michael Nicholson. **Online** maryannekyriakou.de

Wall Pinball, Sintra, Portugal, and Jerusalem, Israel pp.214-215

Completed 2011 (Lumina festival, Sintra; curators Nuno Maya, Carole Purnelle) and 2012 (Light in Jerusalem festival; curator Eduardo Hubsher). **Light Artists** Ocubo (Nuno Maya, Carole Purnelle). **Images** Ocubo. **Online** ocubo.com

Biorama, Montreal, Canada pp.216-217

Completed 2013. **Developer/Venue** Biodôme de Montréal. **Multimedia Artist** Moment Factory. **Manufacture and Installation** Prisme 3. **Equipment and Technical Integration** Solotech. **Photography and Video** Biodôme de Montréal (Claude Lafond). **Graphic Design** Makara (Marc-André Roy). **Music and Sound Effects** Freeworm (Vincent Letellier). **Images** Moment Factory. **Online** momentfactory.com

Qualcomm's Uplinq Launch, San Diego, USA p.217

Completed 2013. **Developers** Qualcomm. **Branding Agency** 1stAveMachine. **Multimedia/Light Artist** Moment Factory. **Music Composer/Director** Freeworm (Vincent Letellier). **VJ** Video Girl (Jennifer Daoust). **Images** Moment Factory. **Online** momentfactory.com

Panorama, Copenhagen, Denmark pp.218-219

Completed 2013 (Strøm electronic music festival; director Frederik Birket-Smith). **Venue** Tietgenkollegiet, Ørestad. **Producer** Andreas Groth Clausen (Frirum). **Light Artist** Jakob Kvist, with Jacob Møller. **3D**

Soundscape Illustrious (Martyn Ware, Asa Bennett). **Music** Mike Sheridan and the Danish National Youth Orchestra. **Lighting Suppliers** [LITE]CO, Martin. **Images** Kim Matthäi Leland. **Online** stromcph.dk

---

Dune, Rotterdam, The Netherlands, and other locations pp.220-21

Completed 2007 (Rotterdam City of Architecture promotion); installations continuing at festivals and venues internationally. **Initial Sponsor** Netherlands Media Art Institute. **Artist** Studio Roosegaarde (Daan Roosegaarde). **Images** Studio Roosegaarde. **Online** studioroosegaarde.net

---

Desire of Codes, Tokyo, Japan p.221

Completed 2010. **Artist** Seiko Mikami. **Producers** Yamaguchi Center for Arts and Media (YCAM), Tama Art University Media Art Lab, NTT InterCommunication Center. **Sound Programming** Satoshi Hama. **Visual Programming** Norimichi Hirakawa. **Technical Director** Soichiro Mihara. **Images** Ryuchi Maruo (YCAM). **Online** idd.tamabi.ac.jp

---

Deep Space, Linz, Austria pp.222-223

Completed 2009. **Venue** Ars Electronica Center. **Design/Production** Ars Electronica Futurelab. **Image** AEC. **Online** aec.at

---

Voyage of Discovery, Wattens, Austria pp.222-223

Completed 2012. **Venue** Swarovski HQ. **Design/Production** Ars Electronica Futurelab. **Digital Pens System** Anoto. **Images** AEC. **Online** aec.at, swarovski.com

---

Amazonas: A Media Opera: Part III, Karlsruhe, Germany, and other locations p.223

Completed 2010 ('Amazone' conference, ZKM Karlsruhe). **Performed with Parts I and II (by others) in** 2010 (Münchener Biennale, Munich, and SESC Pompeia, São Paulo, Brazil). **Production** ZKM. **Concept, Text and Staging** Peter Weibel. **Project Leader** Christiane Riedel. **Music and Sound Design** Ludger Brümmer (ZKM Institute for Music and Acoustics). **Stage Design and Visuals** Bernd Lintermann (ZKM Institute for Visual Media). **Multimedia Stage** Nicholas Völzow, Manuel Weber, Matthias Woelfel, Martin Schmidt. **Dramaturgy, Project Coordination** Julia Gerlach. **Sound Director** Sebastian Schottke. **Realtime Music Environment** Jens Barth. **Assistant Director** Jan Gerigk. **Images** Christina Zartmann, Moritz Büchner. **Online** zkm.de

---

Megaphone, Montreal, Canada pp.224-225

Completed 2013. **Producers** Office National du Film du Canada (Hugues Sweeney), Partenariat du Quartier des spectacles (Pascal Lefebvre). **Designers** Moment Factory (Alexandre Lupien, Étienne Paquette). **Production** Geneviève Forest, Marie-Ève Meilleur, Johanna Marsal. **Voice Recognition System Designer** Computer Research Institute of Montreal. **Images** Moment Factory. **Online** momentfactory.com, megaphonemtl.ca

---

Voice Tunnel, New York, USA p.225

Completed 2013 (Summer Streets celebration). **Sponsor** New York Department of Transportation public art program. **Venue** Park Avenue Tunnel. **Artist** Rafael Lozano-Hemmer. **Custom Hardware and**

Software Stephan Schulz. **AV and Staging** Worldstage. **Production Team** Jordan Parsons, Julie Bourgeois, Karine Charbonneau, Guillaume Tremblay, Claudia Espinosa. **Images** courtesy Rafael Lozano-Hemmer. **Online** lozano-hemmer.com

---

Giant Bicycle Headquarters, Taichung, Taiwan p.226

Completed 2012. **Developer** Giant. **Architect** Studioboase Architects (Ming-Wei Huang, Chih-Jen Wang). **Lighting Design** CMA Lighting Design (Ta-Wei Lin, Chia Ming Liu, Catt Cheng). **Images** Wei-Ming Yuan. **Online** cmalighting.com.tw

---

CYCLE!, Sydney, Australia pp.226-227

Completed 2009 (Vivid festival, Sydney, and repeated as 'Re-CYCLE!' at Vivid 2011). **Light Designers** CLOUSTON Associates. **Technical Advisor** Peter MacLean (Lighting Art + Science). **Bicycle Construction** Jonathon Thwaites. **Image** Jürgen Brinkmann. **Online** clouston.com.au

---

V.I.P., Uppsala, Sweden pp.226-227

Completed 2011. **Developer** Uppsala Municipality. **Venue** Heidenstam School. **Light Artist** Aleksandra Stratimirovi̇́c. **Image** Robin Hayes. **Online** strati.se

---

Fractal Flowers, Céret, France, and other locations pp.228-229

Completed 2014 ('Artificial Paradises' exhibition, Museum of Modern Art, Céret) and earlier exhibitions 2014 (Hong Kong), 2009 (Lisbon, Portugal, and Brasilia, Brazil), and 2008 (Seoul, Korea). **Light Artist** Miguel Chevalier. **Software Programmers** Cyrille Henry, Antoine Villeret. **Technical Producer** Voxels Productions. **Image** Miguel Chevalier. **Online** miguel-chevalier.com

---

Ylem, Lyon, France pp.232-233

Completed 2013 (Fête des lumières). **Developer** Rhône General Council. **Light Artist** Atelier H. Audibert. **Musée des Confluences Architect** Co-op Himmelb(l)au. **Façade Engineers** VS-A Group. **Images** Giacomo Bretzel. **Online** atelierherveaudibert.com

---

Rain Room, London, UK, and New York, USA pp.234-235

Completed 2012 (London), 2013 (New York). **Sponsors** Maxine and Stuart Frankel Foundation for Art. **Light Artist** Random International. **Images** Random International. **Online** random-international.com

---

Magic Carpets, Andria, Italy pp.236-237

Completed 2014 (Festival Internazionale di Andria Castel dei Mondi). **Venue** Castel del Monte. **Light Artist** Miguel Chevalier. **Music** Jacopo Baboni Schilingi. **Production team** Cyrille Henry, Antoine Villeret, Voxels Productions. **Image** courtesy Miguel Chevalier. **Online** miguel-chevalier.com

---

Tapis Magiques, Casablanca, Morocco p.237

Completed 2014. **Sponsors** Casablanca French Institut (Morocco). **Venue** Former Sacré-Cœur Church. **Light Artist** Miguel Chevalier. **Music** Michel Redolfi. **Production team** Cyrille Henry, Antoine Villeret, Voxels Productions. **Images** courtesy Miguel Chevalier. **Online** miguel-chevalier.com

NOVA Display System, Zurich, Switzerland p.238

Completed 2006-2013. **Concept and Visuals** Martina Eberle. **Hardware Development** Supercomputing Systems. **Software Development** Computer Graphics Laboratory ETH-Zurich (Christoph Niederberger, Simon Schubiger-Banz). **Supporters** Blaser Metalbau, ETH-Z Computer Graphics Lab, Genossenschaft Hammer Metallblau, Glas Trösch, Industrial Micro Systems, Swisscom Innovations, Supercomputing Systems, Tribecraft. **Images** courtesy Martina Eberle. **Online** nova.ethz.ch, horao.ch

---

Tom Bradley International Terminal, Los Angeles, USA pp.238-239

Completed 2013. **Developer** Los Angeles World Airports. **Project Director** MRA International. **Architect** Fentress Architects. **Feature Designer/Creative Producer** Sardi Design. **Executive Multimedia Content Producer** Moment Factory. **Content Producer** Digital Kitchen. **Systems Engineer** Electrosonic. **Images** Moment Factory. **Online** momentfactory.com, sardidesign.com

---

Mirrors, London, UK p.240

Completed 2009. **Artist** Random International. **Image** Random International. **Online** random-international.com

---

Oakley Ceiling Screens, New York, USA pp.240-241

Completed 2014. **Developer** Oakley. **Multimedia Artists** Moment Factory. **Digital Fabrication** SITU Fabrication. **Technology Integration** Fulkra. **Image** Moment Factory. **Online** momentfactory.com

---

The Digital Wall, Sydney, Australia pp.240-241

Completed 2013. **Venue** Central Park Retail Centre. **Developer** Fraser Properties and Sekisui House. **Design, Project Management, Content Creation** Ramus Illumination (Bruce Ramus). **LED Supply and Installation** Big Screen Projects (Toby Waley). **Physical Mask and Interactive Kiosk Fabrication** Thomas Creative (David Thomas). **AV Installation** Vision X. **Interactive Games** Current Circus. **Image** Peter Bennetts. **Online** ramus.com.au

---

Skylight II, Stavanger, Norway pp.242-243

Completed 2012. **Venue** New Concert Hall. **Developer** KORO Public Art. **Designer** Inaba. **Engineer** Buro Happold. **Light Designers** Ljusarkitektur, now ÅF Lighting (Kai Piippo, Paul Ehlert, Clara Fraenkel). **Fabrication** DAMSTA. **Animation** MTWTF. **Concert Hall Architect** Ratio Arkitekter. **Images** Ivan Brodey. **Online** af-lighting.com, inaba.us

---

Hylozoic Structures, various locations pp.244-245

Completed 2012 (*Protocell Field*, Dutch Electronic Art Festival, Rotterdam), 2013 (*Epiphyte Chamber*, Museum of Modern and Contemporary Art, Seoul) and 2013 (*Aurora*, Simons store, Edmonton). **Architect/Artist** Philip Beesley Architect Inc. (PBAI). **Studio Team Leaders** Sue Balint, Matthew Chan, Vikrant Dasoar, Faisal Kubba (other production credits available online for each project). **Images** courtesy Philip Beesley Architect Inc. **Online** philipbeesleyarchitect.com, hylozoicground.com

ArRay, Eger and Budapest, Hungary, and other locations p.246

Completed 2012 (Kepes Institute, Eger, and Nomade Gallery, Budapest), 2013 (Digital Shoreditch Festival, London, UK, and Huret & Spector Gallery, Emerson College, Boston, USA), 2014 (Kinetica Art Fair, London). **Artist** Bálint Bolygó. **Image** Bálint Bolygó. **Online** balintbolygo.com

---

Pulsar, Budapest, Hungary, and London, UK pp.246-247

Completed 2009 (Feszek Klub, Budapest), 2010 (Kinetica Art Fair, London, UK), 2012 (La Scatola Art Gallery, London), 2014 (Blank Gallery, London). **Artist** Bálint Bolygó. **Image** Bálint Bolygó. **Online** balintbolygo.com

---

Sky Loupe, London, UK pp.246-247

Completed 2012 (Kinetica Art Fair). **Artist** Bálint Bolygó. **Paving Slabs** Pavegen. **Image** Bálint Bolygó. **Online** balintbolygo.com

---

Aurora, London, UK pp.246-247

Completed 2010. **Venue** Old Town Hall Hotel, Bethnal Green. **Artist** Bálint Bolygó. **Image** Bálint Bolygó. **Online** balintbolygo.com

Selected Reading

Ackermann, Marion, and Dietrich Neumann. 2006. *Luminous Buildings: Architecture of the Night*. Ostfildern: Hatje Cantz.

ALTER - Le Groupe de Bellevue. 1978/1996. 'A Study of a Long-Term Energy Future for France Based on 100% Renewable Energies'. Paris. In *The Yearbook of Renewable Energies 1996/7*. London: James & James.

Armstrong, Rachel. 2015. *Vibrant Architecture: Material Realm as a Codesigner of Living Spaces*. Berlin: De Gruyter Open.

Bazerman, Charles. 2002. *The Languages of Edison's Light*. Cambridge, MA: The MIT Press.

Bowers, Brian. 1998. *Lengthening the Day: A History of Lighting Technology*. Oxford: Oxford University Press.

Brandi, Ulrike, and Christoph Geissmar-Brandi. 2006. *Light for Cities: Lighting Design for Urban Spaces: A Handbook*. Berlin: Birkhäuser.

Butterfield, Jan. 1993. *The Art of Light + Space*. New York: Abbeville Press.

Büttiker, Urs, and Louis I. Kahn. 1993. *Light and Space*. Basel: Birkhäuser.

Carson, Rachel L. 1962. *Silent Spring*. New York: Houghton Mifflin.

Catler-Pelz, Bettina. 2014. 'Taking the Beholder into Account: Looking Back into the Entwined History of Light and Art', unpublished.

Coke, David E., and Alan Borg. 2011. *Vauxhall Gardens: A History*. New Haven: Yale University Press.

Cubitt, Sean. 1993. *Videography: Video Media as Art and Culture*. London: Palgrave Macmillan.

Cubitt, Sean. 1998. *Digital Aesthetics*. Los Angeles, London: Sage.

Cubitt, Sean. 2013. 'Electric Light and Electricity', in *Theory, Culture and Society*. Los Angeles, London: Sage.

Dan, Horace, and E.C. Morgan Willmott. 1907 (republished 2007). *English Shop-Fronts Old and New*. London: Jeremy Mills Publishing.

De Decker, Kris (with Vincent Grosjean, ed.). Undated. 'Moonlight Towers: Light Pollution in the 1800s', in *Low-Tech Magazine*. Online at http://www.lowtechmagazine.com/2009/01/moonlight-towers-light-pollution-in-the-1800s.html

Debord, Guy. 1967. *The Society of the Spectacle*. 2002 edition of Ken Knabb translation: Canberra: Treason Press. Online PDF at http://libcom.org/files/Society%20of%20the%20Spectacle2.pdf

Descottes, Hervé, and Cecilia E. Ramos. 2011. *Architectural Lighting: Designing with Light and Space*. New York: Princeton Architectural Press.

Droege, Peter. 2006. *(The) Renewable City: A Comprehensive Guide to an Urban Revolution*. London: Wiley.

Droege, Peter (ed.). 2008. *Urban Energy Transition: From Fossil Fuels to Renewable Power*. London: Elsevier.

Droege, Peter. 2012. *100% Renewable: Energy Autonomy in Practice*. London: Routledge.

Eco, Umberto (Eng. trans. William Weaver). 1986. *Faith in Fakes: Travels in Hyperreality*. London: Minerva.

Egan, M. David, and Victor W. Olgyay. 2002. *Architectural Lighting*. Boston: McGraw Hill.

European Directorate-General for Communications Networks, Content and Technology. 2013. *Lighting the Cities: Accelerating the Deployment of Innovative Lighting in European Cities*. Brussels: European Commission (Digital Agenda for Europe).

Online at http://ec.europa.eu/information_society/newsroom/cf/dae/itemdetail.cfm?item_id=11175

Falk, David S., Dieter R. Brill and David G. Stork. 1986. *Seeing the Light: Optics in Nature, Photography, Color, Vision and Holography*. New York: Harper and Row.

Fischer, Joachim. 2008. *Light | Licht | Lumière*. Königswinter: H.F. Ullmann/Tandem Verlag.

Franinović, Karmen. 2011 (written 2008). 'Architecting Play', in *AI and Society*, Vol. 26, No. 2, May. London: Springer.

Freeberg, Ernest. 2014. *The Age of Edison: Electric Light and the Invention of Modern America*. London: Penguin.

Gage, John. 1999. *Color and Culture: Practice and Meaning from Antiquity to Abstraction*. Berkeley: University of California Press.

Gardner, Carl, and Raphael Molony. 2001. *Light: Reinterpreting Architecture*. Crans-près-Céligny: RotoVision.

General Electric Company. 1930. *Architecture of the Night*. A Bulletin of the General Electric Company, February. Online at https://archive.org/details/ArchitectureOfTheNightASeriesOfArticlesPublishedByTheGeneral

Haeusler, M. Hank. 2009. *Media Façades*. Ludwigsburg: avedition.

Isenstadt, Sandy, Dietrich Neumann, Margaret Maile Petty (eds.). 2015. *Cities of Light: Two Centuries of Urban Illumination*. London: Routledge.

Jacobs, Jane. 1961. *The Death and Life of Great American Cities*. London: Penguin.

Jacobson, Mark Z., and Mark A. Delucchi. 2009. 'A Plan to Power 100 per cent of the Planet with Renewables', in *Scientific American*, November.

Karcher, Aksel, Martin Krautter, David Kuntzsch, Thomas Schielke, Christoph Steinke and Mariko Takagi (eds.). 2009. *Light Perspectives: Between Culture and Technology*. Lüdenscheid: ERCO.

Koslofsky, Craig. 2011. 'Evening's Empire: A History of the Night in Early Modern Europe', in *New Studies in European History*. Cambridge: Cambridge University Press.

Laganier, Vincent, and Jasmine van der Pol. 2009. *Light and Emotions: Exploring Lighting Cultures: Conversations with Lighting Designers*. Eindhoven: Philips (1st ed.) and Berlin: Birkhäuser (2nd ed. 2011).

Lam, William M.C. 1977. *Perception and Lighting as Formgivers for Architecture*. New York, London: McGraw Hill.

Laposky, Ben. 1958. 'Electronic Abstracts. Art for the Space Age', in *Iowa Academy of Science*, Vol. 65. November.

Lindberg, David C. 1976. *Theories of Vision from al-Kindi to Kepier*. Chicago: University of Chicago Press.

Lowther, Clare, and Sarah Schultz (eds.). 2008. *Bright: Architectural Illumination and Light Installations*. Amsterdam: Frame.

LUCI Association with Jean-Michel Deleuil et al. 2010. *Cities and Light Planning*. Lyon: LUCI.

Lynch, Kevin. 1960. *The Image of the City*. Cambridge, MA: The MIT Press.

McQuire, Scott. 2005. 'Elemental Architectures: Urban Space and Electric Light', in *Space and Culture*, Vol. 8, No. 2. Los Angeles, London: Sage.

McQuire, Scott. 2006. 'Dream Cities: The Uncanny Powers of Electric Light', in John Potts and Edward Scheer (eds.), *Technologies of Magic: A Cultural Study of Ghosts, Machines and the Uncanny*. op. cit.

McQuire, Scott. 2008. *The Media City: Media, Architecture and Urban Space*. Los Angeles, London: Sage.

McQuire, Scott, Meredith Martin and Sabine Niederer. 2009. *Urban Screens Reader*. Amsterdam: NAi/Institute of Network Cultures. (PDF downloadable at http://www.networkcultures.org/_uploads/US_layout_01022010.pdf)

Major, Mark, Jonathan Speirs and Anthony Tischhauser. 2005. *Made of Light: The Art of Light and Architecture*. Basel: Birkhäuser.

Martin, Elizabeth. 1996. *Pamphlet Architecture No 16: Architecture as a Translation of Music*. New York: Princeton Architectural Press.

Michel, Lou. 1996. *Light: The Shape of Space*. New York, London: Van Nostrand Reinhold.

Mitchell, William J. 1996. *City of Bits: Space, Place and the Infobahn*. Cambridge, MA: The MIT Press.

Moyer, Janet Lennox. 1992. *The Landscape Lighting Book*. New York: Chichester Wiley.

Narboni, Roger. 2004. *Lighting the Landscape: Art, Design, Technologies*. Basel: Birkhäuser.

Neumann, Dietrich, and Kermit Swiler Champa. 2002. *Architecture of the Night: The Illuminated Building*. Munich, New York, London: Prestel.

Neumann, Dietrich (ed.), with essays by Robert A.M. Stern, D. Michelle Addington, Sandy Isenstadt, Phyllis Lambert, Margaret Maile Petty and Matthew Tanteri. 2011. *The Structure of Light: Richard Kelly and the Illumination of Modern Architecture*. New Haven, CT, and London: Yale University Press.

Nye, David. 1990. *Electrifying America: Social Meanings of a New Technology*. Cambridge, MA: The MIT Press.

O'Dea, Thomas. 1958. *The Social History of Lighting*. London: Routledge.

Office for Visual Interaction. 2014. *Lighting Design and Process*. New York: Office for Visual Interaction.

Papastergiadis, Nikos, 2009. 'The Paradox of Light: Negative Vision, Being Involved and Imaginative Collaboration', in *X-Blind Spot*. Milan, New York: Edizione Charta.

Papastergiadis, Nikos, Scott McQuire, Xin Gu, Amelia Barikin, Ross Gibson, Audrey Yue, Sun Jung, Cecelia Cmielewski, Soh Yeong Roh and Matt Jones. 2013. 'Mega Screens for Mega Cities', in *Theory, Culture and Society*, December, Vol. 30, Nos. 7-8. Los Angeles, London etc.: Sage.

Perella, Stephen, Laurent-Paul Robert, Vesna Petresin. 2001. 'Hypersurface Architecture: Age of the Electronic Baroque: Studies for a Virtual Campus', in Peter Cook, Neil Spiller and Laura Allen (eds.) *The Paradox of Contemporary Architecture*. London: Wiley.

Petty, Margaret Maile. 2007. 'Illuminating the Glass Box: Architectural Lighting Design and the Performance of Modern Architecture in Post-War America', in *Journal of the Society of Architectural Historians*. Vol. 66, No. 2, June.

Plummer, Henry. 1995. *Light in Japanese Architecture*. Tokyo: Architecture + Urbanism.

Potts, John, and Edward Scheer (eds.). *Technologies of Magic: A Cultural Study of Ghosts, Machines and the Uncanny*. Sydney: Power Publications.

Praun, Sandra, and Aleksandra Stratimirović. 2015. *You Say Light I Think Shadow*. Stockholm: Art and Theory Publishing (Amsterdam: Idea Books).

Ritter, Joachim. 2006. 'The Status Quo of Lighting Masterplans', in *Professional Lighting Design* magazine, December.

Russell, Sage. 2008. *The Architecture of Light: Architectural Lighting Design Concepts and Techniques*. La Jolla, CA: Conceptnine.

Scheer, Hermann. 2011 (English trans.) *The Energy Imperative: 100 Percent Renewable Now*. London: Routledge/Earthscan.

Schielke, Thomas. 2014. 'Urban Data as Light', in *A+U:Data Driven Cities*. Tokyo: JA+U.

Schivelbusch, Wolfgang. 1988. *Disenchanted Night: The Industrialization of Light in the Nineteenth Century*. Berkeley: University of California Press. (First published in German 1983 as *Lichtblicke: Zur Geschichte der künstlichen Helligkeit im 19. Jahrhundert* by Carl Hanser Verlag, Munich, Vienna.)

Seitinger, Susanne. 2010. *Liberated Pixels: Alternative Narratives for Lighting Future Cities*. Cambridge, MA: MIT Media Lab.

Seitinger, Susanne, Markus Canazei and Helmut Schrom-Feiertag. 2012. 'LEDs on the Go: Supporting Pedestrian Flows in Public Transit Networks with Accent Lighting for Increased Efficiencies and Ambient Information Display', in *Proceedings of the 13th International Symposium on the Science and Technology of Lighting (LS13)*, Troy, New York, 24-29 June 2012. Online at Research Gate: http://www.researchgate.net/...Canazei/.../e0b495271083756502.pdf

Shi, Jodi. 2010. *Reducing Artificial Nighttime Light Pollution and Its Impacts*. Stanford: Stanford University Department of Civil and Environmental Engineering report. PDF online at http://www.trianglealumni.org/mcrol/EPA-NNEMS_Light_Pollution_Final.pdf

Tamburini, Fabrizio, et al. 2011. 'Twisting of Light around Rotating Black Holes', in *Nature Physics*, Vol. 7, March.

Venturi, Robert. 1996. *Iconography and Electronics: Upon a Generic Architecture: A View From the Drafting Room*. Cambridge, MA: The MIT Press.

Wands, Bruce. 2006. *Art of the Digital Age*. London: Thames & Hudson.

Weibel, Peter, and Gregor Jansen (eds.). 2006. *Light Art From Artificial Light*. Karlsruhe: ZKM Museum for New Art, and Ostfildern: Hatje Cantz.

Weibel, Peter. 2013. *Rosalie: LightScapes*. Ostfildern: Hatje Cantz.

Whitehead, Randall. 1999. *The Art of Outdoor Lighting: Landscapes with the Beauty of Lighting*. Gloucester, MA: Rockport.

Wigley, Mark. 1998. *Constant's New Babylon: The Hyper-Architecture of Desire*. Rotterdam: 010.

Yot, Richard. 2011. *Light for Visual Artists: Understanding and Using Light in Art and Design*. London: Laurence King Publishing.

Zaidi, Farhan, Joseph Hull, Stuart Peirson, Katharina Wulff, Daniel Aeschbach, Joshua Gooley, George Brainard, Kevin Gregory-Evans, Joseph Rizzo III, Charles Czeisler, Russell Foster, Merrick Moseley and Steven Lockley. 2007. 'Short-Wavelength Light Sensitivity of Circadian, Pupillary, and Visual Awareness in Humans Lacking an Outer Retina', in *Current Biology*, Vol. 17, No. 24, 18 December. Online at http://www.cell.com/current-biology/abstract/S0960-9822%2807%2902273-7

Zarroli, Jim. 2011. 'Tokyo Sees Its Lights Go Dim, and Lifestyles Change'. Online at http://www.npr.org/2011/03/30/134957742/tokyo-sees-its-lights-go-dim-and-lifestyles-change

Index